REFASHIO...

We live in a society as dominated by food preference as by sexual preference, as obsessed with eating too much as with eating too little. Food is the ultimate commodity in an economic system which depends upon the market to meet needs and influences global development and interdependence. Food, from cultivation to consumption, provides the chief link between humankind and the 'natural' environment. Yet, technological advances – in genetics, agribusiness and food processing – have combined with changing patterns of diet and (women's) employment to challenge our perception of the 'natural' and of our position within a 'natural' system. At this point of dislocation, global crisis and conscience over the use we make of the environment have sharpened the ideological force of 'Nature'.

Refashioning Nature analyses the apparently opposed imperatives of the industrial food system and environment. The authors argue that present means of food production, processing and consumption do not satisfy the demands of both North and South, resulting rather in food shortages and surpluses, as well as environmental destruction. One of the major developments within the global food system has been the change in diet associated with the movement of women into the labour market. Beyond the implications for the production of food and the position of the household, this transformation has had a profound effect on the way we manage the environment, and what we assume and perceive is natural.

REFASHIONING NATURE

FOOD, ECOLOGY AND CULTURE

*David Goodman and
Michael Redclift*

London and New York

First published 1991
by Routledge
11 New Fetter Lane, London EC4P 4EE

Simultaneously published in the USA and Canada
by Routledge
a division of Routledge, Chapman and Hall, Inc.
29 West 35th Street, New York, NY 10001

Set in 10/12pt Palatino by
Mews Photosetting, Beckenham, Kent
Printed and bound in Great Britain by
Biddles Ltd, Guildford and King's Lynn

British Library Cataloguing in Publication Data
Goodman, David *1938–*
Refashioning nature : food, ecology and culture.
1. Food. Sociology
I. Title II. Redclift, Michael *1946–*
306.3
ISBN 0-415-06702-2
ISBN 0-415-06703-0 pbk

Library of Congress Cataloging in Publication Data
Goodman, David, 1938–
Refashioning nature : food, ecology, and culture / David Goodman
and Michael Redclift.
p. cm.
Includes bibliographical references and index.
ISBN 0-415-06702-2
1. Food industry and trade – Technological innovations.
2. Agricultural innovations. 3. Food habits. 4. Food supply.
5. Agriculture – Environmental aspects. 6. Food additives – Health
aspects. I. Redclift, M.R. II. Title.
HD9000.5.G66 1991
363.8-dc20   90-27217
CIP

For Annie and Ben

CONTENTS

List of Illustrations   viii
List of Tables   ix
Acknowledgements   x
Introduction   xi

1 FOOD INTO FREEZERS: WOMEN INTO
  FACTORIES   1

2 THE PASSING OF RURAL SOCIETY   47

3 THE ORIGINS OF THE MODERN AGRI-
  FOOD SYSTEM   87

4 INTERNATIONALIZATION AND THE
  THIRD WORLD FOOD CRISIS   133

5 ENGINEERING LIFE: AGRI-
  BIOTECHNOLOGIES AND THE FOOD
  SYSTEM   167

6 THE FOOD SYSTEM AND THE
  ENVIRONMENT   201

7 CONCLUSION: COUNTER REVOLUTION   239

Bibliography   257
Index   272

ILLUSTRATIONS

FIGURES

2.1 Owner-occupation of farms in the UK   63
3.1 Acreage, yield and production of corn 1870–1960,
    US   106
3.2 Agriculture: labour, capital and total factor
    productivity   113
3.3 World price trends of major agricultural
    commodities   125

PLATES

1.1 Women working in food processing, wine
    cooperative, Hungary   12
1.2 Women working in food processing, wine
    cooperative, Hungary   13
2.1 French charcuterie displaying non-industrially
    processed food   66
2.2 Spanish peasant family farm, Galicia, Spain   67
2.3 Modern irrigation in Castille, Spain   71
4.1 Indian women buying food from a government
    CONASUPO store, Mexico   138
4.2 Farming systems researcher conducting trials on
    small farm, Guatemala   151

TABLES

1.1 Appliance ownership 1964-87   20
1.2 Refrigeration combinations   20
1.3 Microwave ovens - ownership growth   21
1.4 The freezer-microwave link   21
1.5 Employment in the food system 1987   27
1.6 UK grocery retail market shares 1987-8   41
1.7 Historical margins in grocery retailing   41
1.8 Percentage share of total grocers - long
    term - UK   42
2.1 Concentration of landholding in the UK 1982   65
4.1 World grain trade 1934-80   153
6.1 Intermediate consumption in French agriculture   205
6.2 UK countryside recreation 1985   224
6.3 Membership of environmental organizations in
    the United States   224
6.4 Agriculture and the environment:
    an international perspective   227

ACKNOWLEDGEMENTS

In writing this book, we have benefited greatly from discussions with friends and colleagues, notably Fred Buttel, Judy Clark, Bill Friedland, Mauro Borges Lemos, Jan Pahl, Clive Potter, Nanneke Redclift, Shigeo Shiki, José Eli da Veiga, Sarah Whatmore, Bill Adams and John Wilkinson. Their comments and suggestions undoubtedly have improved the presentation of our arguments, although the usual disclaimers apply. We are very grateful to Janet Roddy and Marjorie Smith for their help in preparing the manuscript for publication.

David Goodman
Michael Redclift

INTRODUCTION

Contemporary interest in food is not confined to pleasure in its consumption, but extends in every direction: to its economic importance, the semiotics of food taste, the dangers of food additives and the politics of food security. We live in societies as dominated by food preferences as by sexual preferences, as obsessed about eating too little as by eating too much. In addition our interest in food is associated, for good and evil, with our interest in 'nature'. As we begin to become aware that we are in a position to destroy our environment, for the first and last time, 'nature' has become imbued with so many virtues that the term 'natural' no longers confers unambiguous meaning. Nature commands attention, and the 'natural' has an ideological force, which takes us to the heart of the paradox of development itself. We have refashioned nature, in our minds, as well as in test tubes and fields, transforming ecological processes into political axioms. We are poised on the threshold of a very dramatic breakthrough in thinking; we can no longer evade the global biospheric imperatives, such as climate change, without at the same time closely scrutinizing the products of science and technology which transform every-day life. And at the centre of the paradox is our view of nature, biological and social, in the modern food system.

The conventional approach to the food system is to concentrate on its component parts: geographical, cultural and conceptual. We possess, as a consequence, a rich literature on agricultural development, technology, food policy (both national and international) and diet. Since the 1970s we have also seen a burgeoning literature on farming, 'organic' and chemically-based, and on biotechnology. The environment has kept our attention too, and ecological problems have begun to be clearly associated with social and economic problems and policies. It is difficult to talk about food, at least in the industrialized North, without also talking about the environment. Similarly, we are learning, to our peril, that even global environmental problems often boil down to local concerns – not least with the ways in which, and the terms under which, food is produced and consumed. There is no end to the debates, and no foreseeable end to the problems. Each of the areas we point to involves a story and an explanation, and we experience our greatest difficulties when we are forced to put these explanations together and make coherent sense of them.

We are attempting in this book to formulate an integrated framework of how the modern food system developed, and how this system works. At a structural level this means grappling with the connections between technology and policy, patterns of consumption and production, social justice and the directions that structural change is taking. This much we see as political economy: a sometimes unpopular approach, which is needed now more than ever. However, we are also seeking to address more intangible matters that are less amenable to structural analysis, such as the shifting definition of gender roles associated with food, and the patterns of resistance to environmental change and the elaboration of the modern food system. Our analysis is pitched at the level of systemic interconnections, whether local or international, rather than linear, historical causation. We seek to argue that the modern food system developed around structurally compatible processes, changes in technology and the labour process, inside and outside the home, but in ways that have themselves changed our view of the relationship between food and nature. As we observe in Chapter 6, the modern food system has epistemological importance, both across cultures and within them, which we ignore to our cost.

The structure of the book requires some explanation before we can examine the argument that is being unfolded. Chapter 1 ('Food into freezers: women into factories') is about the social processes which have accompanied the increasing commoditization of food, singling out the British experience during this century as a focus for our attention. The argument is that the diversification of household consumption, especially since the Second World War, has coincided with changes in the labour process, which have drawn more women into paid employment outside the home. Specifically, the production of consumer durable goods for the home ('white goods'), and more processed foods for consumption, is linked to changing patterns of work based on gender relations. Women's labour, like food, has been only partially commoditized, and food preparation in the home has acquired enhanced market value from greater investment in its production, and from media attention to the 'naturalness' of women's role within the domestic context. The social impulse of the food system within the advanced, industrial societies has been to shift ideological attention inwards from the community to the private household, while women's labour has been drawn outwards into the wider industrial labour process. The economic impulse which parallels these social changes is observed in product differentiation, lengthening the food production process, and 'adding value' to food, to ensure that profitability is maintained in an otherwise shrinking economic sector. It can be argued, then, that the modern food system represents a new construction of both social and economic domains, of the private (household) as well as the public (community), a shift with an economic logic located in accumulation (Chapter 3) and in the penetration of overseas markets (Chapter 4).

The shifting distinctions between the household and the wider society are taken up, within the context of agricultural production, in Chapter 2 ('The passing of rural society') where the focus of analysis, again, is the United Kingdom and Western Europe. This chapter explores the way in which family farming has gained in political legitimacy in Europe since the Second World War while, simultaneously, there has been a further breakdown in the 'natural' organic cycles that traditionally defined farming activities. The relationship between the sources of livelihood, for farming families, and the changes in farming practices are explored at the farm and local community level. In order for the modern food system to develop, it is suggested that changes in the agricultural sector were required which have increasingly separated food production from sustainable farm husbandry. The requirements of the modern food system include both a reliable and stable supply of food and the social stability necessary to bring about increases in production. The 'end of rural society' has been marked by the development of an agricultural sector more closely linked with industry than with the rural environment, in which increased political legitimacy was traded for the traditional role of environmental stewardship. Some of the costs of these processes are explored later in the book. For example, the implications of the breakdown in natural organic cycles for the growth of environmental resistance and how this has contributed to rising awareness of the dangers of modern food consumption are looked at in Chapter 6 and the Conclusion.

The main thrust of the argument in Chapter 2 is that the 'social contract' between farmers and the state, in the industrialized countries, brought a measured social peace to rural areas, but at the same time facilitated the greater integration of agriculture and industrial activities. A relatively stable family farming structure acquired greater legitimacy from state support and more diversified livelihoods, leaving the wider society to grapple with many of the consequences, such as the environmental impacts. Family farming was both the means to achieve agricultural restructuring within the wider food system and, through encouraging the enhanced participation of farmers in the political process, a way of reducing resistance to the implications of refashioning nature *on the farm*, including greater concentration in agriculture, the rural exodus and pluriactivity.

In Chapter 3 ('The origins of the modern agri-food system'), we examine the emergence and institutional development of the modern food system against the background of the transition of industrial capitalism from extensive to intensive or 'fordist' accumulation. A central theme is that the biological production/consumption cycle underpinning the food system has determined highly specific patterns of accumulation and industrialization. These patterns are explored using the concepts of 'appropriation' and 'substitution', tracing the industrialization process from early mechanization and the rise of agri-chemicals to the revolutionary advances in plant genetics incorporated in hybrid seeds. This analysis also questions the view that this process signifies the extension of 'fordist' industrial organization to agriculture, although such parallels are apposite in downstream manufacturing and catering sectors, notably fast-food operations. Institutional developments in the food system are seen as the outcome of the drive of industrial capitals to control 'nature' in agricultural production and food manufacturing and the interest of the state in pursuing cheap food policies.

Since 1930, this articulation has created distinctive state–agriculture relations which are represented as a new technology/policy model, whose main elements are considered in detail. This model, first institutionalized by New Deal farm support programmes in the United States and later extended to other industrial economies, unleashed the enormous production potential of modern agriculture. Production surpluses and food aid policies provided the foundations of the post-war international food regime under US hegemony, which brought unprecedented stability to world commodity markets in the 1950s and 1960s. These institutional questions are examined at some length in Chapter 3, together with the growing contradictions of the 'productionist' model – farm concentration, environmental degradation, budgetary strains and farm trade disputes – which came to the fore in the 1980s. The legitimacy of agricultural production subsidies, including the environmental 'stewardship' role traditionally claimed by farmers, has increasingly been challenged, drawing new constituencies into the politics of the food system.

Chapter 4 ('Internationalization and the Third World food crisis') examines the origins of the crisis now found in many Third World countries. The discussion locates the loss of food security in the specific trajectory taken by the international diffusion of the US technology/policy model and consumption norms discussed in Chapter 3. Although this transfer is constrained by structural distortions, the South has been exposed to the contradictions of the modern food system developed in the North. The diffusion process initially was determined by the coincidence of interest linking US crop surpluses, Cold War politics and the implementation of cheap food/cheap labour strategies of accelerated industrialization in the Third World countries. Peasant agriculture and rural self-provisioning have been chronically weakened by cheap food policies, aggravating rural poverty and shifting urban consumption patterns towards imported grains and animal protein. The dependence of Third World countries on imported food, which is such a distinguishing feature of the new international division of labour, in turn enhances the strategic importance of export agriculture, further displacing staple food crops and their peasant producers.

These issues are examined using case material, drawn mainly from Latin America, on agricultural development policy and the impact of transnational corporations on Third World food systems. This discussion emphasizes the asymmetry between agricultural supply structures and domestic food requirements, which is attributed, in part, to the *de facto* internationalization of the agricultural resource base in the South. Closer integration in international export markets is achieved by rapidly depleting natural resource 'capital', as in the case of Central America's tropical forests, while the pressures exerted by the rural poor on marginal ecosystems are intensified. The food and environmental crises now gripping Third World countries have common roots, it is argued, in the distortions and contradictions of the modern food system.

Chapter 5 ('Engineering life') examines the revolutionary implications of biotechnologies for the restructuring of the modern food system. The genetic code now can be manipulated for commercial gain, opening up the prospect of refashioning nature according to the logic of the market place. This chapter explores the emerging technological base of the food system, new modes of integration between agriculture and industry, and the forces influencing the deployment of agri-biotechnologies. Research and policy agendas increasingly are dominated by large private corporate actors as governments embrace the chauvinistic rhetoric of global technological rivalry. The privatization of agri-genetics research has potentially far-reaching consequences for the Third World, including impacts on agricultural export markets, farming systems and bio-diversity. These issues are considered by looking at current efforts to strengthen intellectual property rights to living organisms, North–South confrontation over plant genetic resources, and channels of technology transfer. The closing discussion of the environmental release of genetically engineered organisms emphasizes how biotechnologies will radically transform the relationship between farming and the environment as market selection, rather than natural selection, becomes definitively enthroned as the guiding principle of the modern food system.

Chapter 6 examines some of the costs of the modern food system, especially through price support and technological innovation, and notably the loss of sustainability. Increasingly, management interventions are required to contain the destructive effects of modern agriculture, to ensure the better conservation of the environment, and to handle the social resistance which relatively rapid environmental change has brought about. The chapter explores the political, economic and cultural dimensions of resistance to the modern food system, and suggests some ways in which conflicts have been contained in developing countries. The paradox with which we began thus comes to mind once again: sustainability begins to be valued as it is being lost, and nature takes on greater importance the less it is preserved. Having broken into 'natural' systems, relocating activities previously undertaken on the farm to other parts of the food system, the direction of the transformation is away from sustainability towards increased industrial integration, while the concerns of the political process are focused on ways of managing the contradiction. To ensure that the cycle of renewal is not completely destroyed, the intensification of agriculture is deliberately arrested (conservation areas, organic farming, wholefoods) as the refashioning of nature meets the social resistance of groups (in both developed and developing countries) whose own consciousness it has served to transform.

Nature is being refashioned in several ways, and the relationship between these processes is not usually recognized. Recent changes in agriculture are closely linked with technological developments in industry and with the wider context of capital accumulation in advanced industrial countries. The discussion of this wider context through reference to regulation theory, at the beginning of Chapter 3, may prove demanding for readers lacking specialized background in the social sciences. However, the argument is summarized later in Chapter 4. The pattern that emerges from these pages is of a close and mutually reinforcing relationship between human ingenuity in refashioning nature and the process of social and economic restructuring that are necessarily involved. Without changes in the labour process and in gender roles, the radical transformation of agriculture would not be reflected in the wider food system. Without changes in agriculture and food consumption habits, the 'natural' environment would not have been subjected to such rapid transformation. In refashioning nature we are doing no more, and no less, than refashioning ourselves.

1

FOOD INTO FREEZERS:
WOMEN INTO FACTORIES

The modern food system developed around several components, each of which is complex in itself, and usually interpreted by specialists within the social and natural sciences. For example, technological changes in agriculture (Chapter 3) and food manufacture (Chapter 7) are the province of economists and food technologists, and though they have attracted a lot of attention, agricultural change and the transformation of food through industrial processes are rarely linked to the social and economic processes which help to explain both of them. They are treated as discrete areas, as separate, if related, concerns. In a similar way the discussion of the urban household's food consumption, and the changes in the work undertaken by women within this household, although it has attracted the interest of feminist scholars, is not usually linked to the changes occurring within the household farm economy. In this case the urban–rural divide serves to divorce two closely related social and economic processes, which have divested women of direct (technological) control over the transformation of organic nature and food. The shifts in women's labour market participation, which have accompanied the development of the modern diet in urban areas, have come about partly because of the kind of food we eat, and the availability of domestic technologies. They also mirror what has happened on the farm; in most of the advanced industrial societies women play a larger role in the family farm, and have assumed roles which might have been performed in a more specialized salaried form, had the family farm (usually owner-occupied) not been the dominant social institution in rural areas.

1

As we shall see in Chapter 2, the transformation of the family farm in turn reflects changes in the wider industrial economy, notably in the supply of industrial inputs to farming, which echo those within the 'downstream' food processing industries. It is essential, then, that we view the food system *as a system*, and the connections between parts of this system as essential to its formation and development. These components – women's labour in the urban household and on the farm, technological change in agriculture and in food processing, ecological systems viewed both as resources for agriculture and necessary to the preservation of other values – are usually addressed as discrete realities, within relatively bounded disciplinary traditions. In our view this makes more urgent the need to break the boundaries of thinking, of intellectual categories. The analytical (and empirical) connections between the labour process, technology and policy are not easily summarized, and it would be facile to do so without giving proper attention to the specific issues involved at each stage of the argument. Nevertheless, much of the hard work in exploring these connections must lie with the reader – and with future researchers. Where there is a shift of gear between chapters, or between different levels of analysis, we trust the reader will bear with us, and recognize our attempt to signpost some of the problems, as well as some of the benefits, of a very wide-reaching approach. This book often considers familiar territory in an unfamiliar context, and one much broader than is usually considered appropriate. In taking the discussion through successive chapters, and successive components of the argument, we invite the reader to consider connections and explanations which are sometimes tentative but always potentially illuminating. To view the whole picture we need to begin with the parts; but, unlike much of the literature we shall be discussing, we do not intend to end with the parts, merely to suggest that forging new connections between ideas and issues carries its own dangers as well as benefits.

This chapter examines the role of domestic labour and changes in food consumption in the development of the modern food system. Women play a particularly important part in most aspects of food production, processing and consumption, in developed and developing countries. In developing countries there is a considerable literature which draws attention to this role, and it would not be an exaggeration to state that our understanding of rural development has been transformed by our understanding of women's work and status. In the industrialized countries the picture is rather more patchy; perhaps it is testimony to the advance of the modern food system that much less research and policy criticism has been devoted to women's role in the food system of the North, than that of the South. Recent work in the United Kingdom and the United States has pointed to the way in which housework has been transformed during this century (Gardiner, Himmelweit and Mackintosh 1980, Bose 1982, Meissner *et al*. 1988). Important research has been published on the allocation of money within the family (Pahl 1983, 1989). Similarly, empirical research on the role of women's management of food, within the family, was undertaken during the 1980s (Charles and Kerr 1988). However, we still lack a body of literature which addresses the central question: what part have changes in gender divisions played in the way food is prepared and consumed within the family?

This chapter addresses this question, but, in the absence of studies whose principal focus is the gender dimension of the food system, we can only begin to answer some of the larger, empirical and theoretical issues. It is nevertheless important to raise them at the outset, and to flag their importance for further research and understanding. Gender, like class, needs to be theorized as a category of social power relations, with an active structuring potential of its own. Questions such as how skills become gendered, and how divisions in the labour force shape gender identity, are complex issues, which require close attention to historical evidence and to the role of ideology at an intimate, personal, level, within the household and in defining 'the family'. We shall argue that the economic and technological processes which mark the development of the food system rest upon changes not only in human behaviour but also in human perceptions. The analysis of these perceptions should encompass the ideology of food preparation and consumption, but current knowledge barely enables us to understand the shifts in the perception of women's roles on the broader canvas. We know next to nothing about how changes at the ideological level have influenced or facilitated the development of the modern food system.

It is clear that women's roles have been radically transformed during the last century, in all the industrialized countries. Every social group, from servants to single parents, and most social phenomena of industrialized societies, from child labour to class divisions in women's work experience, indicate major changes in the structure and internal relations of domestic food consumption. One concomitant of industrialization has been a significant shift in the way in which food is produced, processed and consumed, especially in the division between the home and the factory, the two principal locations where food is processed. Similarly, as we shall see in the next chapter, women play a central role in the transformation of food production, through changes within the household farm economy. Again, to fully appreciate the part played by gender relations in these changes we need to unravel a largely unwritten and often invisible history: that of food practices within the household, and their relationship to food ideologies in the society at large.

This chapter explores some important facets of the question, but, inevitably, it only begins to explore some of the most important issues. It would be useful, for example, to distinguish between households, on class and ethnic lines, in exploring the balance between domestic labour and paid labour for women in food processing. Similarly, the whole area of food self-provisioning in the evolution of modern industrial society needs to be opened up to more rigorous analysis. If we examine working-class households over time, then it is clear that women in such households have long possessed extended working lives outside the confines of the household. However, we still have only the vaguest, most impressionistic sense of how these experiences have shaped tastes and patterns of food consumption, and habits of food preparation in the home. Some of the social changes surrounding discussion of the family in industrial society today, such as the 'new servant class' of child-minders and nannies, or the effects of prolonged male unemployment on the domestic division of labour, have clear implications for the future evolution of the food system. As we shall show in the concluding chapter, the advertising and marketing of food products and domestic

technology today makes great use of different household consumption styles, and the food industry has adopted different
marketing strategies for what are perceived as different
segments of the market. However, although we know more
about the food people consume, we know very little about
changes in the tasks and responsibilities that govern the preparation of food, beyond the fact that they are largely in the hands
of women.

A central concern, it will be argued, is that of causation.
To what extent have changes in the household, and in the role
of women, influenced or helped to bring about wider changes
in the food system? By the same token we can ask: how have
changes in the wider food system influenced or helped to bring
about changes in gender divisions within the household?
Clearly both processes are important and interrelated, but to
approach the question through a crude theory of causation
would be unhelpful. As much of the work connected with food
(like other work) has turned into 'jobs', so female employment
has become 'packaged' into different activities, located in different parts of the food system. This would not have been possible
without technological changes or, as we shall see in Chapter
3, an economic system based on increasingly intensive accumulation. As the purchasing power of most consumers has risen,
so the transformation of food has engendered a transformation in work. Labour and food have long been commodities,
but their interrelationship has changed over time, as the locus
of consumption has moved further from the locus of production. Both also provide evidence that what we look upon as
'natural' is in fact infinitely variable; food has been subjected
to naturalization, it has been socially constructed, but changes
in consumption habits are also closely linked to technological
and economic changes. In place of simple cause and effect, we
suggest a model with component parts each able to influence,
and transform, one another. Rather than emphasize causative
links between ideology, social roles and technical change,
we suggest thinking in terms of a system. In a sense, our point
of entry into this system could be through production (the
farm), consumption (the household) or processing (the food
industry). By considering the point of consumption first,
we are simply placing the emphasis with the universal,
biological necessity which food conveys, and which societies

5

shape: the need to eat. We return to consumption in the concluding chapter.

To appreciate the key role played by shifts in family consumption and women's employment in the way in which the food system has evolved, we need to examine the historical process through which a rural society that largely consisted of rural food producers became one largely made up of food consumers. The Industrial Revolution, and the development of a large, urban working class, marked the beginning of the process through which food, and the labour which went into producing it, became fully commoditized.

The principal clue to these changes is provided by Engels in *The Conditions of the Working Class in England in 1844*, in which he discusses the way in which a growing urban population pressed on limited natural resources, forcing the 'agricultural revolution' to confront increasingly *extensive* technological limits:

> in consequence of the increase of population, the demand
> for agricultural products increased in such measure that
> from 1760 to 1834 some 6,840,540 acres of waste land were
> reclaimed; and, in spite of this, England was transformed
> from a grain exporting to a grain importing country.
> (Engels 1892: 13)

Urban population increase in this period was a necessary stimulus to industrialization (although historians continue to debate the relationship between cause and effect) because industry produced many of the commodities required by the enlarged domestic market and, especially as the nineteenth century advanced, by overseas markets as well. At the same time, food was the single most important wage-good in the newly industrialized society, and food needed to be imported if enough of it could not be produced domestically. It was essential to the development of manufacturing industry that food was acquired cheaply, just as the raw materials for the textile industry had to be acquired cheaply. Engels noted that, by mid-century, the 'reduction of the cost of raw material, and the keeping down – if not as yet the bringing down – of wages' remained the principal objective of manufacturing capital (Engels 1892: xii). By the late nineteenth century an international food order had been created, linking Britain at its hub to the newly created 'settler societies' in North America and

Australasia, where grains and meat were being produced for the market provided by the core industrial countries.

At the level of macro political economy this process is discussed more fully in the next chapter. At the moment it is sufficient to note that the shift from a cheap food policy, under early industrialization, to the development of a wider consumer market, not simply for food or other basic commodities but for a range of sophisticated consumer goods, domestic technology or 'white' goods, played a critical role in the development of late capitalist industrialization.

The food consumed by the household has increasingly, in the past half-century, become processed and prepared outside the home, in the food manufacturing sector, service industries and, through a growing sophistication in household food preparation, using new domestic technologies. As we shall see in Chapter 2, the full elaboration of the food system involves important changes at the farm level, which in some respects parallels changes in food manufacturing and within urban households. To begin to understand the full importance of the modern food system, we need to consider the transformations taking place in the 'parts' themselves: the family and women's labour in urban areas; the consumption of food; and the technology of food production, processing and manufacture.

Our starting point in this chapter is women's labour, including food preparation, in the urban working-class household.

WOMEN'S WORK: THE TRANSFER OF DOMESTIC SKILLS

Until the First World War the British middle and upper classes relied heavily on large numbers of domestic servants to clean their houses, wait on them at table and prepare their food. This area of women's domestic experience – there were many more women servants than men – reproduced in other people's homes what women were called on to do in their own. By 1921 there were 1,232,000 domestic servants, only 80,000 fewer than before the war, and the demand from the middle class for servants was increasing (Dawes 1984: 164). After the war, unemployment had risen again, and economic hardship had forced many former domestic servants back into service. Even full-blown economic depression, however, could not undo the changes in women's expectations that the war had introduced. In 1901, one in three girls between the ages of 15 and 20 was a domestic servant, and 28 per cent of the female servant population was less than 20 years old. After the First World War the shortage of servants became an officially recognized 'problem', and led in 1924 to a Report from the Central Committee on Women's Training and Employment, which argued for shifting the costs of training a new generation of 'skivvies' to the taxpayer. It was argued at the time that domestic training was not only useful in servants to the middle classes, it was also essential in working-class wives (Dawes 1984). This echoed the way in which, during the nineteenth century, the professions concerned with diet, as they became the preserve of women, were progressively marginalized from mainstream medicine (Belasco 1989: 198). Women's domestic role was 'professionalized' in both the middle and working classes.

In the 1930s, women's work in most urban working-class households was extremely onerous. The classic description of Marjorie Spring Rice captures the unremitting toil which women were forced to undertake. It also serves as a benchmark from which to observe many of the changes introduced by the modern food system during the last half century:

> When once she is up there is no rest at all till after dinner. She is on her legs the whole time. *She has to get her husband off to work,* the children washed, dressed and fed and sent to school. If she has a large family, even if she has only the average family . . . , *four or five children,* she lives . . . in a house extremely inadequately fitted for her needs. Her washing up will not only therefore be heavy, but it may have to be done under the worst conditions. She may have to go down (or up) two or three flights of stairs *to get her water,* and again to empty it away. She may have to heat it on the open fire, and she may have to be looking after the baby and toddler at the same time. When this is done, she must clean the house. If she has the average family, the rooms are very 'full of beds', and this will make her cleaning much more difficult than if she had twice the number of rooms with half the amount of furniture in each. *She lacks the utensils* too; and lacking

8

any means to get hot water except by the kettle on the fire, she will be as careful as possible not to waste a drop. The schoolchildren will be back for their dinner soon after 12, so she must begin her cooking in good time. Great difficulties confront her here. *She has not got more than one or two saucepans and a frying-pan,* and so even if she is fortunate in having some proper sort of cooking stove, it is impossible to cook a dinner as it should be cooked, slowly and with the vegetables separately; hence the ubiquitous stew, with or without the remains of the Sunday meat according to the day of the week. *She has nowhere to store food,* or if there is cupboard room, it is inevitably in the only living-room and probably next to the fireplace. Conditions may be so bad in this respect that *she must go out in the middle of her morning's work to buy for dinner.* This has the advantage of giving her and the baby a breath of fresh air during the morning; otherwise, unless there is a garden or yard, the baby, like herself, is penned up in the nine-foot-square kitchen during the whole morning.

Dinner may last from 12 till 3. Her husband or a child at work may have quite different hours from the school-children, and it is quite usual to hear this comment. Very often she does not sit down herself to meals. The serving of five or six other people demands so much jumping up and down that she finds it easier to take her meals standing. If she is nursing a baby she will sit down for that, and in this way 'gets more rest'. She does this after the children have returned to school. Sometimes *the heat and stuffiness of the kitchen* in which she has spent most or all of her morning takes her off her food, and she does not feel inclined to eat at all, or only a bite when the others have all finished and gone away. Then comes the same process of washing up, only a little more difficult because dinner is a greasier meal than breakfast. After that, with luck at 2 or 2.30 but sometimes much later, if dinner for any reason has had to go on longer, she can tidy herself up and *rest, or go out, or sit down.*

(Rice 1939: 89; emphasis added)

Within this summary description of women's domestic work,

9

and especially their roles in food preparation, cooking and attending to family needs, we find a number of points with which this chapter will be concerned. Housework, including food-related work, was not simply tedious and repetitive for working-class women, it was also dirty, cramped and exhausting. Women performed the labour of buying, preparing and serving food, among other tasks. In addition, compared with the 1930s few households today have to obtain their water in buckets, and the lack of basic cooking utensils used in the home no longer constitutes a major problem for the 'housewife'. Food storage is better and the reorganization of food retailing and the widespread use of convenience foods have transformed the time allocated by women (and men) to provisioning for the family. There is considerable discussion surrounding the 'drudgery' of housework today, and the fact that it is not shared equally by men. There can be few women, however, who would, if given the choice, opt for the full-time domestic work endured in a pre-war working-class family trapped by poverty. There are broadly two views on the effect of industrial food processing on women's labour, both in the home and outside it. The first view argues that developments in technology, and changes in women's employment, have been largely beneficial to women. This is the *conventional* view:

> the development of food manufacturing has a significance which transcends purely nutritional questions, for it is clear that manufactured goods have played an important part in transforming the lives of a great section of the British population. The essence of food manufacture is that, by removing the burden of food preparation and cooking from the home to the factory, it allows more time for the pursuit of other more desirable activities. *This has been an important factor in the liberation of women during the Twentieth Century,* but it has also affected the lives of others, young and old alike, for whom the preparation of food represents an arduous task.
>
> (Johnston 1977: 67; emphasis added)

Other writers go still further, emphasizing that the changes in women's work, accompanying the development of industrial food manufacture, represent new *choices* for women:

10

in the United Kingdom the high proportion of working women influences the demand for convenience foods, especially of good quality, since *multiple earner* families often[take home] a higher income in total. The combination of greater affluence and work commitments implies a desire to spend less time in shopping, preparing and washing-up meals, and enhances the demand for one-stop shopping trips, labour saving kitchen appliances and food that is easily managed.

(Burns 1983: 8)

Some of the implications of changes in household size for diet and food consumption are dealt with later in this chapter.

There can be little doubt that for most women the advent of new household technologies has taken some of the burden out of housework, although as we shall see, this has not led men to share the burden that remains. However, what is noticeable from the quotations is the almost casual way in which the enormous increase in women's waged work is commented on. Changes in domestic food preparation, among other things, may have released women for more paid employment but they have not put paid to housework. Indeed, as Charles and Kerr (1988) show, more paid employment has not materially affected women's primary responsibility for preparing food. Clearly, too, changes in food consumption, including the increase in 'convenience foods', probably would not have taken place at all without women playing the role of second wage earner. Given their obligations to their job outside the home, it is obvious that women could not have undertaken the traditional tasks of food preparation and cooking in the home. In short, it was not an accident that the labour market for women, and food consumption habits both inside and outside the home, changed in the way they did. The two processes are interconnected and interdependent. Together they imply a considerable shift in women's roles and responsibilities, these have changed together with the process of accumulation in modern capitalist society which, among other things, has shifted the burden of work for women.

The principal motor of change, from the standpoint of the household, has been the transformation of women's working roles which, within the space of thirty years, has led a majority of women to find employment outside the home. Women's work in the wider economy is concentrated in relatively few occupations, and these closely parallel the kind of work women have traditionally performed within the household: cleaning, catering and performing personal services. There is now a rich vein of literature dealing with women's labour in the home, much of it espousing a very different perspective from what we termed the *conventional* approach above. The main thrust of the *feminist* view of housework is that it has subjected women to social and psychological pressures, while in financial terms women's contribution to the social reproduction of the household has been undervalued (Beneria and Stimpson 1987; Yeandle 1984; Westwood 1984; Malos 1980; Sharpe 1984; Whitelegg *et al.* 1982). Among the most thoughtful of these accounts, and one which places emphasis on the impact of new domestic technologies or 'white goods', is that of Bose (1982) writing about modern North American households.

The transfer of women's labour from food preparation in the home to the processing industries is sometimes compared with the similar shift in textile production in nineteenth-century

Plate 1.1 Women working in food processing, wine cooperative, Hungary

Britain. As manufacturing took over from cottage industry, women left the domestic handloom, and their labour was transferred to the mill:

> Women had always been involved in the family production of textiles. When textile production was removed to the factories, girls entered the factory workforce. But, even in that setting, they were seen as working for the household. In Italy and France, some factory owners tried to create 'family' conditions and supervision for their female employees and, on occasion, even to arrange marriages for them.
>
> (O'Day 1985: 43)

Plate 1.2 Women working in food processing, wine cooperative, Hungary

The transfer of many of women's skills, and much of their labour, from the household to food processing was in some respects a similar process, occurring at a later stage in the industrialization of the United Kingdom, but there were also significant differences. Food processing in industry did not remove all aspects of food preparation from the household nor, more significantly, did it remove from women the key role as family providers, whose primary responsibility it was to try to ensure adequate family nutrition.

HOUSEWORK AND FOOD PREPARATION: THE DOMESTIC GHETTO OR THE NEW LIBERATION?

Women still bear the responsibility for most food preparation and cooking in the home. They also do most of the work involved in these activities. As we shall see, the introduction of new technologies into the home, as well as women's working lives outside it, have served to accelerate the development of complementary features of the food system. Not only is it impossible to see the wages coming into the home primarily in terms of the male 'breadwinner's' interface with the formal economy, it also prevents us from understanding the transformations that have occurred outside the home.

Another way of approaching the issue of women's labour in domestic food preparation, and housework generally, is to concentrate on the gradual reduction of what Huws (1985) calls 'unsocialised labour', women's unpaid work for the family and community, in contrast to a bigger role in wage employment. The focus then becomes not the household but the process through which unsocialized labour is transferred into the market:

> The abolition of household production often reduces disposable family incomes, as additional goods and services have to be purchased. It also reduces any last traces of women's economic independence . . . thus, by a combination of push and pull factors, women are drawn into waged labour.
>
> (McDowell 1989: 175)

This approach is very close to the one we have adopted, which places emphasis on the essential interdependence of changes in the labour market and changes in the production and consumption of food commodities. However, it is not clear that there is what McDowell calls a reduction in disposable family incomes in this context. McDowell seems to be referring to the distinction between real and monetary income, where the former includes use values, as well as exchange values. Certainly the self-sufficiency of the family declines, in the sense

14

that it directly produces less of what it consumes, but disposable *family* income usually increases as women enter the labour force in larger numbers. The important corollary, of course (which McDowell refers to), is that this process has an inexorable momentum of its own, as part of the drive within capitalism to provide, or invent, more and more goods for the mass market to consume.

In examining whether the acquisition of 'labour-saving' white goods reduces women's domestic labour we are clearly dealing with ideological, as well as economic, issues. Yeandle (1984) argues that whether or not these items are labour-saving, women *believe* them to be. In many cases they are considered essential items.

> Whilst washing machines, vacuum cleaners and refrigerators were seen by most women as basic essentials which were required in every home, some of the less commonly held items were seen as aids to working mothers which they wanted, but were only able to afford because they were in paid employment.
>
> (Yeandle 1984: 165)

Yeandle adds that the 'alternative strategy' open to women, that of reducing their workload in the house by lowering the standard of household care (as recommended most provocatively by Conran 1977), is open to the criticism that it seriously neglects the ideological basis of these standards. Yeandle is arguing that standards of household care assume more, rather than less, importance for women who are not full-time housewives.

Another perspective on the division of labour between the household and the formal economy is provided by writers like Pahl (1984), Gershuny (1985) and Mingione (1985), who seek to explain the variation that occurs in most societies around the production of use values for the household, and wage employment in the formal economy. Mingione notes that the way households manage social reproduction is not mechanistically determined by capital, nor is it a function of some blind technological imperative. Social reproduction at the level of households, including housework and food preparation, also depends 'on pre-existing, socio-historical conditions and, to a certain extent, on a degree of choice among different options

15

by the individual unit of reproduction – the household' (Mingione 1985: 18). He cites as an example the fact that some households continue to process or grow food for self-consumption, even when it is no longer attractive from a strictly economic point of view. In similar vein, in the same volume of essays, Cornuel and Duriez (1985) show how the owner-ship of allotments in Northern France enabled families to provide vegetables for their own consumption, of superior quality to those available in the market place. People still grow vegetables, although the opportunity cost is quite high, because vegetables that are bought in the shops are looked upon as 'second best'. Clearly there is considerable variation, and important cultural preferences, in determining how much effort is put into different household activities. In view of research findings like these on the 'informal economy' in developed countries, the most profitable approach to women's outside employment and the demands of housework, might seek to recognize explicitly both the ability of dual-earning households in making strategic moves to maximize their advantages as consumers, and the interest of the industrial process in encouraging the transfer, or substitution, of female labour. The possibility exists of what Mingione (1985) describes as a 'complementary relationship' between specific forms of social reproduction and the kind of labour that industry requires for its further development.

THE ACQUISITION AND USE OF DOMESTIC TECHNOLOGY

In considering the allocation of time and effort between the domestic economy and the formal sector of employment we need to pay particular attention to the role of 'white goods' technologies which, as we have seen, have ideological as well as practical importance in defining household activities. Of key importance in the feminist literature has been the question of whether women exert more control over the technology they employ in the home than the technology they use in the factory. Cynthia Cockburn (1985) argues that women's earning capacity has led to a neglect of their traditional domestic skills, partly because these skills require more time than women can afford, and partly because they seldom

16

generate enough income to make them as attractive as out-side employment. However, Cockburn maintains, women are not in control of the technology in their homes either:

> Normally, women use utensils and implements – the dishwasher, vacuum cleaner, car. They do not use tools. The utensils and implements are, in their way, tools, of course, and they are used by women with skills (making food, sewing clothes) certainly equal to the male skills of their husbands. But women cannot fix these utensils and implements when they go wrong. It is men on the whole who are in control of women's domestic machinery and domestic environment.
>
> (Cockburn 1985: 220)

On this reading, white goods have done little to alter the divisions of labour through which male dominance is maintained in the home. This argument does not address the question which remains central to the development of the food system. This is that the introduction of new technology to the home, and women's working lives outside it, have enabled changes to take place in other components of the food system, notably agriculture and the food processing industries, which *have* served to alter the content of people's lives in radical ways. Women preparing the Sunday lunch may well be trapped within the patriarchal family, as Westwood (1984: 170) argues, whatever appliances were used to produce it, but it remains important to consider how the use of those appliances, and the technology that goes into the food, have contributed to the patterns of work and consumption that lie at the heart of the family. Charles and Kerr are examples of sociologists who, in addressing the place of food within family life, barely consider the way food technologies have affected behaviour or the way family life is structured (Charles and Kerr 1988).

One of the more searching analyses of the role of technology within the home in reshaping women's roles, and the construction of housework, is that of Bose (1982). Bose poses the question, what is more liberating: less physical or mental exertion, cheaper appliances or less time spent on housework? Drawing on evidence that the amount of *time* spent on housework has not declined, she argues that to some extent 'the only true liberation from housework comes when someone else does the work' (Bose 1982: 227). This, manifestly, has not happened, although she also notes that women in the paid labour force spend an average of 3 to 4 hours a day less on housework than full-time housewives. Bose does not attach much significance to this reduction in housework, placing her emphasis, instead, on the opportunities household appliances provide for women to foresake full-time housework for wage labour outside the home. She distinguishes (after Hartmann 1974) between three major types of household technology: *utilities* (such as running water, electricity, gas, and sewage and refuse collection services); *appliances* (the machines that do the housework) and *commodities* (notably processed foods). In Bose's view, utilities have had the greatest impact on the time-consuming, heavy labour of housework. However, she feels that the time saved was probably transferred to other household activities, rather than 'saved' for the wife. She quotes evidence from the United States that the time allocated to childcare, shopping and general household management has increased, while meal preparation time has decreased.

Bose also reiterates the view that time saved by the use of white goods in the home has led to the purchase of more goods of this sort, rather than an expansion of leisure time. Once initiated, the conversion to domestic technology seems to provide a momentum of its own, as consumers seek more sophisticated appliances, or increase their wage-earning time in an attempt to acquire more expensive durable consumer goods. Echoing Westwood's description of the English Sunday lunch, Bose argues strongly for the *symbolic* importance of white goods, since the new technology can help women legitimize their position as good nurturers (this, after all, is the moral emphasized by advertisers). Household appliances also serve to increase the social status of the family, a factor which probably bears on men's consciousness of their value, too. In Bose's view, the increased use of white goods in the home has served to 'privatize' women's work, taking it out of a communal context and locating it firmly in the household. Advertisers have sought to emphasize the responsibilities of women for their families, making housework 'so emotion-laden that moving it into a more communalized term may be impossible' (Bose 1982: 233).

Finally, Bose argues that 'improving working conditions in the home is the opposite of freeing women for other tasks'

(1982: 234). In her view women have grown increasingly dissatisfied with housework as other opportunities within society have opened up. New household technologies have served to 'reinforce the home system, keeping women marginal to the larger society' (1982: 234), they have enabled standards of housework to rise, but they have not equalized work within the family; men still undertake a relatively small amount of work in the home, compared with their wives. A re-division of labour among household members has not occurred, but household appliances, particularly appliances for food preparation, have cut down the amount of time spent on this work, and extended women's role in the home, encouraging a stereotyped division of labour.

A close examination of the expansion in 'white goods' within the home suggests that they have played a considerable role in 'freeing' women from many of the more important chores in housework, although it is clear that the gender division of labour has not been substantially altered, and it may well be true that women are subjected to increasing ideological pressure to provide higher standards in house cleanliness and efficiency. It is clear that in recent years the acquisition of domestic technology has played a major role in redefining housework, and making the environment in which (mainly) women work more pleasant. We have only to reflect on the description of women's domestic work in the 1930s provided by Rice (1939) to appreciate the difference. Since 1964 British homes have acquired washing machines (87 per cent), fridges (56 per cent), fridge-freezers (44 per cent), freezers (37 per cent) and, most recently, microwave ovens (36 per cent) during a period when homes with central heating rose from 13 per cent to 75 per cent, and home ownership from 45 per cent to 65 per cent (see Jenkins 1988 and Tables 1.1 to 1.3).

These appliances were major consumption items, absorbing significant amounts of household income and making the home environment more comfortable. The addition of refrigerators, freezers and microwave ovens also enabled food to be stored for longer, and for convenience foods to be used much more easily. The proportion of fully fitted kitchens has increased to almost one-third of households, and kitchen extensions approach almost 200,000 a year (Jenkins 1988: 255). Survey research shows clearly that most women regard a partially fitted kitchen, of adequate size, and containing a refrigerator, freezer, microwave and food mixer, as 'essential' to the proper performance of domestic work. In 1963, when the AGB home audit was established to measure the market for consumer durables, the extraordinary growth in their ownership was not predicted.

The 'gentle revolutions' which have seen the widespread improvement in food storage and convenience has now been almost eclipsed by the rapid conversion to microwave cookery. Within the space of five years microwave ownership increased

Table 1.1 Appliance ownership 1964–87

	1964 %	1969 %	1974 %	1979 %	1984 %	1987 %
Washing machine	54	64	70	75	83	87
Clothes dryer	17	25	30	33	42	45
Dishwasher	1	1	2	3	5	8
Fridge	35	59	77	75	66	56
Fridge-freezer	n/a	n/a	3	15	31	44
Separate freezer	n/a	n/a	10	25	35	37
Gas cooker	62	61	57	56	51	51
Electric cooker	33	37	43	41	49	49
Microwave oven	n/a	n/a	n/a	n/a	11	36
Central heating	13	28	46	56	66	75
Home ownership	45	49	53	55	61	65

Source: Jenkins (1988)

Table 1.2 Refrigeration combinations

	Total homes %	AB %	C1 %	C2 %	D %	E %
Fridge only	23	13	18	17	23	47
Fridge-freezer only	37	30	38	39	44	30
Freezer only	–	–	–	–	–	–
Fridge + fridge-freezer	1	3	1	1	2	1
Fridge + freezer	31	43	35	36	23	14
Fridge-freezer + freezer	5	9	6	5	4	3
All three	1	2	–	1	–	–
None of these	1	–	–	–	–	4

Source: Jenkins (1988)

Table 1.3 Microwave ovens – ownership growth

	1985* %	1987* %
Total homes	18	36
Social class		
AB	24	46
C1	20	40
C2	22	43
DE	9	21
Age of housewife		
16–24	14	33
25–34	22	45
35–44	26	47
45–54	24	45
55–64	17	31
65+	5	16
Size of household		
1	6	15
2	16	33
3	22	43
4	26	50
5+	24	49

Source: Jenkins (1988)
*As at December 31

Table 1.4 The freezer–microwave link

	% homes	% owning microwave oven
Total homes	100	36
Fridge only	23	11
Fridge-freezer only	37	36
Freezer only	–	44
Fridge + fridge-freezer	1	39
Fridge + freezer	31	51
Fridge-freezer + freezer	5	56
All three appliances	1	65
No refrigeration appliances	1	5

Source: Jenkins (1988)

from 11 per cent to about 40 per cent, reflecting important shifts in diet and food tastes. The quick adoption of microwaves is linked to several factors, including the availability of cheaper, Far-Eastern sourced products, which now account for an increasingly high share of the market, an increasingly competitive retail trade 'anxious to get behind any new product development to lever additional profit' (Jenkins 1988: 257) and a food industry determined not to suffer from a decline in the proportion of higher incomes spent directly on food.

It is also clear that freezer volume and microwave ownership are linked (see Table 1.4), and mark new food preparation and serving practices. White goods ownership can be seen as fulfilling essentially complementary functions, linked to saving time and increasing convenience. An indication of the linkages between white goods acquisition in apparently different areas of domestic life is signalled by the rise in microwave ownership after 1984/5, at a time when the sales of video recorders were beginning to level out. It is possible – even probable – that given the relatively wide dissemination of VCRs some households turned their attention to the time they could save (to watch the video) by investing in a microwave. From a marketing standpoint, at least, it is worth emphasising that microwaves do not standardize food itself, they standardize food *variety*, opening up a market for microwave foods as varied as the market for microwaves. The very real sociological significance of white goods, especially in food consumption, although largely ignored by sociologists is well described in this passage, written by a market researcher:

> Microwave ovens are important in changing eating and cooking habits, but there are a host of other products which contribute to the housewife's food preparation armoury . . . Think of the slow pervasive growth in ownership of salad spinners, wine bottle re-corkers . . . or even some of the tiny items you keep in your cutlery drawer. We have a device for de-stringing runner beans, which is perfectly designed for the job and worth its weight in gold. Then there are more complicated pieces of equipment such as food processors, mixers, coffee makers, sandwich toasters, slow cookers, deep fat

fryers, pressure cookers, all of which have had an influence on the way we eat.

(Jenkins 1988: 258)

A serious consideration of the increased use of white goods in the home leads to a number of conclusions. First, it is clear that housework, especially food preparation, is lighter work than it was for most working-class women. Indeed, although food preparation in the home is still essentially boring work, it may be no more boring than the work undertaken by the working-class mothers and grandmothers of young women today. Second, it is difficult to maintain (as Bose does) that domestic food technology helps to confine women to a role in the home, when to acquire this technology women (especially) need to earn an outside income. The development of white goods makes housework less onerous, although the boredom often induced fails to gain the understanding of men, and does not force them to share more household tasks. These are certainly strong points in the feminist case on housework. However, it is possible to argue that work in the home today is more culturally laden, and that women are subject to ideological manipulation in maintaining these cultural definitions, without this necessarily implying a greater degree of 'domestication' of women's labour (compared with the past) or that the framework in which it occurs is more exploitative.

In our view the present situation is put succinctly by Meissner *et al.* (1988) in the conclusion to their study of housework in Canada:

> Paid work offers to married women the potential of at least some financial independence from their husbands but, at the same time, confirms their domestic dependency in the menial and subordinate character of their paid work. The larger contradiction, according to which employment demand and domestic requirements accumulate in their dependent labour, leaves no exit for wives.
>
> (Meissner *et al.* 1988: 493)

EMPLOYMENT IN THE FOOD INDUSTRY: AN EXIT FOR WIVES?

At the outbreak of the Second World War the British

people spent nearly £1.5 billion on food, almost a third of the National Income. The industrialization of food was already well advanced. Writing during wartime, George Darling recognized that

> the food trades have expanded from a business partner-ship between farmer, merchant and shopkeeper, which aimed at providing the simple means of sustaining life, into a large collection of related industries some of which are still concerned with satisfying our primary wants, but most of which are organized to produce and sell luxuries.
>
> (Darling 1941: 87–8)

Most of the food for the food industries – two thirds of it – came from abroad, and most of these imports were of food staples, including wheat, butter and cooking fats. The war-time Ministry of Food took over control of food imports, almost £600 million a year by 1940, making it the largest single business operation in the world, at that time. The coalition government had drafted the mandarins of the private food industry into government service, as Commodity Food Controllers, and from the onset of hostilities Lord Woolton was advised by, among others, thirteen managing directors and chairmen of public companies involved in food manufac-turing, processing and retailing.

The wartime objective, to produce more food at home and to ration food more fairly, was to have considerable import-ance for future relations between the food industry and the state. It was to produce the conditions under which more women could work outside the home, in food preparation as well as war-related industries, and government gradually came to take some responsibility (although grudgingly) for improved standards of nutrition for the masses. Wartime reorganization was also the key factor in encouraging more people to eat out-side the home: mass catering had arrived.

By 1941, communal feeding centres, named 'British Restaurants' by Lord Woolton, had opened in more than 100 towns. There were over 150 of them in London alone. At most of these restaurants a nutritionally adequate (if not always appetizing) meal could be bought very cheaply. Some of the restaurants also served meals on a self-service basis, something quite new to the British consumer. These restaurants clearly met a need in wartime, but also provided the stimulus for post-war thinking; women were working outside the home, and people had more cash to spend on food. In post-war Britain, workers increasingly ate their midday meals in the works canteen, an institution which was given impetus by wartime conditions. The expansion of outside catering answered the call of economic independence and consumer preference. In many respects wartime conditions continued after the War, but the development of the restaurant, catering and food manufacturing industries really gained further impetus from the early years of enforced shortages.

The post-war story built on wartime experience:

Subsequently the need for ready-cooked meals was kept alive. Founded upon labour-intensive production techniques, the new service jobs necessitated the employment of a flexible reserve army of workers (whose labour power could be utilized and shed according to the rate of labour-saving technological development) in order to secure the fastest rate of capital accumulation.

(Baxter and Raw 1988: 63)

Between 1951 and 1981 women rose from under one-third of the paid labour force to roughly a half, with most of this increase, as we have seen, in the service sector. This was also the period in which 'eating out', especially what we call 'fast food' today, became important. For example, during this period the United Kingdom drew into the vortex of its developing service industries not just women from the domestic sphere – many of whom had worked outside the home for the first time during the war – but also men and women from the colonies. An example is that of Chinese catering. Chinese take-aways were the outcome of social and demographic changes during the 1960s and 1970s: immigration from Hong Kong combined with new sources of income and an appetite for 'exotic' foods. The expansion of the food industry helped to make household names of many foreign-owned companies, and these took advantage of a transfusion of labour from overseas. This labour came to be supplied by people displaced from the agricultural labour process in the colonies:

Rapid industrialization in Hong Kong [had] outstripped the food supplies produced by the rural hinterlands,

25

the New Territories. Agriculture was transformed into cash cropping of specialized rice grains and vegetables, whilst the main bulk of the colony's rice was imported cheaply from Thailand. This led to the swift demise of previous forms of rice production as a viable source of income throughout the New Territories and the hastened disintegration of the local economy, as pressure for industrial, commercial and housing land encroached into rural areas.

(Baxter and Raw 1988: 63)

The failure of local economies in the global periphery assisted the diversification of the market for food in the centre. Chinese men began to leave the colony in larger numbers for the United Kingdom. Their forefathers had come to Britain as seamen; they came as personal contacts allowed, to work in the restaurant and catering industries. During the 1960s most Chinese immigrants worked in the restaurant side of the business, but later – just as they had been pushed abroad by domestic restructuring – they were pushed 'sideways' by competition from fast-food chains like Kentucky Fried Chicken and McDonald's. The falling profitability of family-run restaurants led Chinese immigrants towards smaller capital units operating on lower running costs. By 'the mid- to late 1970s . . . the simultaneous decline of Chinese restaurants and the rapid spread of Chinese take-aways and fish and chip shops' (Baxter and Raw 1988) had taken place, within a radically restructured fast-food market.

Today, employment in the food industries accounts for many more people than employment in agriculture, forestry and fishing combined (see Table 1.5). In June 1987 there were more than half a million people employed in food, drink and tobacco manufacturing in the United Kingdom, almost twice the figure for the 'primary producing' sector. Another quarter of a million people worked in food distribution, half a million more in food retailing, and finally over a million additional workers in hotel and catering. A total of 2.8 million people worked within the food system of this country, excluding, of course, home-based food activities. The technological transformations behind these changes are examined in Chapter 3.

26

Table 1.5 Employment in the food system 1987

Sector	Total employed	Part-time	Women	Proportion women (%)
Agriculture*	300,200	57,800	82,300	27
Food manufacturing	544,800	90,000	224,100	41
Food distribution	252,000	41,100	85,500	35
Food retailing	595,400	316,500	377,400	63
Hotel and catering	1,094,700	630,100	728,500	67
TOTAL	2,787,100	1,135,500	1,497,800	54

Sources: *Employment Gazette*, and the Department of Employment, see Covey 1987.
*Including forestry and fishing

Of these 2.8 million workers 41 per cent were part-time, almost twice the national average (Covey 1987). Among part-time workers, not surprisingly, women predominated; women outnumber men by three to one in the food industry as a whole, while in some sectors, like food retailing and the hotel and catering industry, the vast majority of workers are female (see Table 1.5). A recent survey by the largest trade union in the industry, the General, Municipal and Boilerworkers (GMB), revealed that there is considerable bias against women in labour agreements throughout the food industry. Women workers suffered from job segregation, locating them in 'ghettos'; they were systematically excluded from bonus schemes and pension rights; few had maternity leave agreements; and pay differentials between the lowest graded jobs (mainly performed by women) and basic craft rates (mainly men) were often impossible to justify (Covey 1987). The research showed that pay differentials between men and women within the industry were explained by two factors: women worked fewer total hours and were confined to the lowest graded jobs, giving them less access to bonuses and shift-pay.

During the last decade there have been several changes in the food industry which have been able to exploit specific features of women's employment. First, there has been an increase in part-time employment. In 1981 64 per cent of workers in the industry were full-time, but by 1987 this figure had fallen to 59 per cent. In some cases firms have replaced full-time workers by part-timers, as a matter of policy. Second, employment in farming and food manufacturing had fallen, by 1987, from 37 to 31 per cent, but catering and retail employment had actually increased. Technological innovations have enabled food retailers to gear their employment practices to the peaks and troughs in customer buying, through mechanisms like Electronic Point of Sale (EPOS) which has computerized stock accounting for major retail chains. Accordingly, staff time can be made to 'fit' busy shopping periods – making it much easier to use part-time workers who work relatively short hours. Similar technological changes in fast-food retailing have also led to more part-time workers on very short shifts (and work contracts). In 1986 a survey by the National Economic Development Office (NEDO) found that 94 per cent of food and drink manufacturers were seeking to increase the flexibility of their workforce; the trend towards using women on a short-term basis, to supplement full-time male workers, and to facilitate the introduction of new technology, is clearly set to continue.

The data in Table 1.5 also suggest another feature of the developing food system that requires comment. Since the 1960s it is retailing, rather than food manufacture, that has been the dominant influence in the food system. As we have seen, women have been drawn into the wage-labour force through a process of restructuring which is linked with changes in food consumption. These changes in consumption are linked, in turn, to a shift from the production of use values in the home to the production of exchange values outside the home.

As the importance of the consumer has increased, so the food industry has concentrated its efforts to maximize profits in the sectors of the industry which show most growth potential. Food products have become part of wider marketing and advertising strategies, pursued in the face of a declining share of family consumption of food. These trends have occurred at a time of growing capital concentration in the food industry. By 1980, six 'buying points' (major food retailers) accounted for over 40 per cent of grocery sales in the United Kingdom. This concentration at the consumer 'end' has given retailers considerable power over food manufacture, and even food production in the fields.

Given the impossibility of increasing the volume of food consumed, food manufacturers have had to concentrate on high-value foods, through increasing the length of the food production process. Food products become more 'convenient' and more expensive, and are given an extended shelf-life through growing sophistication in food preservation. These food products are also, as Lang and Wiggins (1985) show, composed of a narrower range of primary agricultural products, notably wheat, rice and soya. The pressure from retailers to develop new product lines – 'product differentiation' – has led to shifts in manufacture, which, in turn, have fed back to the agricultural producers, many of them from developing countries.

In countries like Brazil and Thailand increased monocultivation of land is linked, albeit indirectly, to changes in food consumption practices in the United Kingdom and other industrialized countries. Monocultivation is an element in the ecological crisis afflicting developed and developing countries alike.

We find evidence from changes in food manufacturing, then, of fundamental shifts in the labour process of developed countries, of important changes in food consumption and diet in these countries, and, finally, of major changes in land use and trade on the part of developing countries, whose patterns of food consumption and ecological systems are also subject to transformation.

FAMILY FOOD CONSUMPTION: CURRENT PRACTICE

The evidence of contemporary accounts shows that before the modern food system became fully established this century most people in Britain spent nearly 60 per cent of their income on food, about the same proportion as in most Third World countries today (Oddy 1976). Agricultural labourers in Kent spent nearly 70 per cent of their wages on food and drink in the early nineteenth century (Richardson 1976). During the decade 1900 to 1910 the percentage most households spent on food was probably slightly more than 60 per cent. By 1955 household expenditure on food, as a percentage of total consumers' expenditure, had fallen to only 27.5 per cent. By 1980 it was only 17.2 per cent. In the period between 1961 and 1985, as expenditure on food continued to plummet as a proportion of total expenditure, most households in the United Kingdom *doubled* their expenditure on other important items such as housing, and all forms of transport (Halsey 1987). It is interesting to note, however, that where women control the intra-household economy, a higher proportion of household income is spent on food (Pahl 1989).

As food came to make up a smaller part of household expenditure, processed food became more important. According to Frank and Wheelock (1988), processed food represented over 70 per cent of household food expenditure by 1980. Before the Second World War most of the food consumed had been imported into the United Kingdom; by 1980 most of it was home-produced. Even at the outbreak of the Second World War, Britain was still importing 87 per cent of its cereals, 73 per cent of sugar, 69 per cent of cheese, 51 per cent of meat and bacon, 39 per cent of eggs, 25 per cent of vegetables and 92 per cent of all fats. As the British Empire was disbanded, so cheap food from the colonies took second place to home-produced (and heavily subsidized) food, a process which was really given most impetus by the Second World War, when 'digging for victory' was a strategic part of the war effort. In the 1920s and 1930s many forms of processed food – especially canned foods – were largely imported. Domestic canning increased in importance during the 1930s, but even in 1938, before the outbreak of war, 70 per cent of the canned food consumed in the United Kingdom was imported. Britain was the world's leading importer of canned foods.

The food people consume has not only changed as a proportion of their total expenditure as the standard of living has improved, it has also changed in content. The farm labourer with whom Howard Newby lived in Suffolk in the early 1970s consumed an insignificant amount of the food his labour had been used to produce, but he possessed a deep freeze. The relative isolation, both geographically and socially, ensured that farm labourers and their families used a deep freeze 'to cover the long periods between major shopping trips' (Newby 1983: 35). A deep freeze was a necessity, forced on such workers by changes in retailing and, more indirectly, by the specialization of most farm production itself.

Before 1900, food consumption surveys show that not only was a very high proportion of disposable income spent on food, much of it was spent on meat, which was much more expensive than it is today, in real terms. Poorer people spent a higher *proportion* of their meagre income on meat than those who were slightly better off. Among the poorest families, over one-third of food expenditure was on meat, although quantities of meat consumed were usually pitifully small. Meat was used to flavour an extraordinarily boring diet, which consisted of a few components, mainly carbohydrates and fats. Bread and potatoes were the staples, supplemented by some (animal) fat and sugar or treacle. Suet puddings and treacle, together with dumplings flavoured with meat stock, brought a little relief from the tedium of 'filling' but insipid starches. Women and children ate much less meat than men; urban working-class families ate fresh vegetables and fruit on the rarest of occasions (Oddy 1976).

According to Margaret Reeves, the weekly menu of a working-class family, with the man in employment, was seriously inadequate in nutritional terms:

> To boil a neck [of mutton] with pot herbs on Sunday, and make a stew of 'pieces' on Wednesday, often finishes all that has to be done with meat. The intermediate dinners will ring the changes on cold neck, suet pudding, perhaps fried fish or cheap sausages, and rice or potatoes. Breakfast and tea, with the exception of the husband's rashers, consist of tea, and bread spread with butter, jam or marg.
> (Reeves 1913: 111)

Meal patterns, before the First World War, centred around the presence or absence of the male 'breadwinner' (as they often do today) and women and children not only ate less, they frequently ate at different times from the man. Children passed the day tugging on a piece of bread, spread with dripping or a scraping of jam. The advantage of bread for feeding children was that it could be eaten without a plate or fork. Much sentimental reflection surrounds family mealtimes in popular mythology, but for most urban working-class families mealtimes were not the popular feasts they are sometimes depicted as being (Rowntree 1901).

Women's diets within the family were every bit as bad as that of the children, and in many cases they were worse. Rowntree remarks that:

31

we see that many a labourer, who has a wife and three or four children, is healthy and a good worker, although he earns only a pound a week. What we do not see is that in order to give him enough food, mother and children habitually go short, for the mother knows that all depends on the wages of the husband.

(Rowntree 1901: 135)

In a sense, *intra-household* food consumption was as unequal in the early 1900s as *inter-household* consumption. As more women have entered the paid labour force so women and children have tended to become better nourished (also a trend observed in the Third World today). The 'price' of these improvements has been to extend women's working responsibilities, the ubiquitous 'double day', and to place a growing emphasis on housework standards and the 'need' for an ever-increasing number of domestic appliances, aimed at 'reducing' the boredom and exertion of housework. It is difficult, nevertheless, to imagine women's and children's levels of nutrition improving as they did, especially during and after the Second World War, had women not been earning the wages to pay for better food.

In 1938, when most British food was still imported, and food made up a sizeable proportion of total household expenditure, over a quarter of the British population was offically judged malnourished. Even among those not in the poorest quarter of the population, many 'could not afford to buy enough of the right kinds of food to keep themselves physically fit' (Darling 1941: 15). Sir John Orr estimated at the time that almost half the population spent less than ten shillings per week on food, and only 13 million of a total population of 46 million spent enough money on food to obtain an adequate diet. In another survey, conducted in 1937, over half the families with four children or more lived below the poverty line (Darling 1941). It was estimated that even in 1941, when rationing was beginning to 'spread' less food more equally throughout the population, there were still 7 million people below the minimum nutritional standard.

One way of measuring the effect of changes in the real price of food is to consider how long it takes for a household to work to gain specific food items. Halsey (1987) calculates that, in the period, between 1961 and 1985, when household food consumption dropped from over a third of total expenditure to under one fifth, the length of time necessary to work to pay for a loaf of bread fell from 9 minutes (in 1971) to 4 minutes (in 1985) for dual-income families. Comparable figures for other items show the same trend: the 22 minutes necessary to pay for a dozen eggs in 1971 had become 8 minutes by 1985. In addition, this trend held for single-income households too, although not to as marked an extent (Halsey 1987). Many other important items of household expenditure, such as electricity, petrol, coal, car licence and cinema admission, needed increased work-time by 1985.

The same report shows that, as expected, low-income families allocate more of their income to food: 29 per cent of their expenditure went on food, compared with only 15 per cent of the expenditure of high-income households (1987: 108). Food also lost its importance within family budgets in other European countries, such as West Germany, the Netherlands, Belgium and Italy, during the 1970s (Gershuny 1985: 134). In the period between the early 1970s and the mid-1980s there was also considerable expenditure on new 'white goods', many of them linked to food. Between 1973 and 1984 the percentage of households with children in the United Kingdom who possessed a refrigerator rose from 86 to 97 per cent; those with a deep freeze (once considered a 'luxury' item) rose from 52 to 75 per cent in the same period. This increased dependence on food that could be stored, in an earlier period canned rather than frozen, was the summation of longer-term trends, through which processed food had entered the popular diet.

FOOD COMMODITIES AND THE PROCESSING INDUSTRIES: A POPULAR DIET IN THE MAKING

Food processing was already well developed in Britain by the 1880s, but it was not until much later, really the 1950s, that this was fully reflected in the working-class diet. As we have seen, the diet of most urban, working-class families was an impoverished one in the first decades of this century. What effect did changes in the food industry, and employment in the economy as a whole, have on this pattern?

Some of the more important changes in food consumption had their greatest impact on the middle classes, but gradually found their way into the diet of much poorer families. Two examples of food commodities of this type are biscuits and breakfast cereals. They also provide a good example of the enhanced power of advertising and the impact of the new 'multiple' grocers (and, much later, supermarkets) on working-class eating habits.

Biscuits had originally been consumed largely by sailors: the ship's biscuit had developed with the scale of Britain's maritime interests, and it was staple fare for seamen throughout the world in the eighteenth and nineteenth centuries. By 1833, ship's biscuits were manufactured by steam-driven machinery. Biscuits were a rather dull but necessary part of the diet of servicemen (including soldiers after the Boer War at the end of the century). But by the 1860s, Huntley and Palmer possessed twenty factories, and with the development of multiple grocers in the 1870s and 1880s – such as Lipton, Home and Colonial and the Co-operatives – biscuits soon became part of middle-class eating habits (Corley 1976). The growth of demand for biscuits paralleled that for tea and sugar, which increased fivefold between 1840 and 1900. Products like biscuits and cakes were almost unknown to the working-class diet until after the Second World War, but they had entered mass consumption much earlier, and even influenced the time of day when the middle classes took their meals.

Changes in cereal consumption, similarly, spread down the social pyramid after they had first been tested, and approved, by the middle class. Packeted breakfast cereals were very much a marketing innovation of the early part of the century, but they did not reach the mass market in the United Kingdom until after the Second World War. In fact, cereal consumption in 1900 was almost twice that of today, at 237 lbs per capita, but whereas most cereal was consumed as porridge at the turn of the century, the advent of packaged breakfast cereals changed the *form* in which they were consumed (Collins 1976). The real marketing and packaging innovations in the breakfast cereal market were developed in the United States, where the wax-tight container and advertising aimed at the children's market, had a major influence on other sections of the food industry. In the 1930s in Britain, writes Collins, 'a typical

34

breakfast in the lowest income group consisted of bread and dripping, with a little bacon at week-ends' (1976: 36). Surveys of eating habits show that ready-cooked cereals, rather than cooked breakfasts, became progressively more important in the late 1950s, 1960s and 1970s. This was also a period in which the ownership of the breakfast cereal part of the industry was concentrated in fewer hands (Belasco 1989: 188). Gradually, in the view of the public, bread was becoming an 'inferior good' and breakfast cereals were considered a 'superior' one. In fact, bread consumption continued to be buoyant throughout this century, but most families continued to buy the white wheaten loaf, rather than wholemeal or brown bread. The white loaf, which had accounted for 95 per cent of bread consumption in 1900, still accounted for over 80 per cent of bread consumed, by weight, in 1970. The major innovation, at the time of breakfast cereal expansion, was the introduction of pre-sliced bread.

Another food item which has played a major part in the processing industry this century, as well as in the popular diet, is refined sugar. The consumption of refined sugar increased from 50 lbs per capita in 1874 to over 100 lbs in 1960, most of this increased consumption occurring before the First World War. An important factor in the increased consumption of sugar was the decline in its price; after sugar beet became widely available early this century, the *real* price of sugar declined dramatically, compared with other food items, like cheese, butter, eggs and honey. According to Johnstone (1976), by 1970 over half the sugar consumed in the United Kingdom was used by the food manufacturing and processing industries, mainly in the form of soft drinks, cakes and confectionery.

Perhaps one of the most indicative items in the working-class diet, and an interesting barometer of wider social mores, is tea. The per capita consumption of tea doubled between 1870 and 1910, leaving other drinks like chocolate and coffee far behind. It has sometimes been suggested that this phenomenal increase in tea consumption was linked to the activities of the temperance movement in Britain in the late nineteenth century. Whether or not more tea was consumed *because* of temperance activity remains obscure, but it is clear that, at least for a couple of decades, increased tea consumption closely paralleled the decline in alcohol consumption per head of population.

From the 1890s onwards, the fortunes of tea, pre-eminently a 'colonial' product, and one which played a critical role in working-class diet, were linked with the food system generally:

> The pattern of the new mass retailing was integral with tea-selling, but had been set slightly earlier with other commodities, particularly ham and bacon. The characteristics were the same: the shops sold a very few lines in a mass trade, offering good quality at low margins, taking advantage of economies of scale and the rapidly falling prices available from imported commodities, advertising and aggressively selling, to tap the extra purchasing power flowing to the urban working class.
>
> (Mathias 1976)

This new pattern of market expansion for basic food goods, linked closely with their packaging and processing, was the principal retail revolution before sophisticated household technology could take the 'convenience foods' market a step further in the 1960s. It has been argued, and quite convincingly, that before the First World War the rate of growth of food technology probably restricted the introduction of new forms of food. As Oddy (1976) makes clear, the technological application of the new science of bacteriology together with changes in refrigeration, began to affect baking, meat, fish preservation and dairy produce during the decade up to 1914. The development of a mass market for most processed food, introducing new *forms* of food, was associated with women's entry, in vast numbers, into the paid labour force, and the mass production of white goods. Recent, and emerging, trends are important in this respect.

RECENT CHANGES IN FOOD CONSUMPTION: CONVENIENCE AND THE DUAL-INCOME HOUSEHOLD

Total consumer expenditure on food today is a much smaller proportion of total household expenditure. In 1986, at current prices, it was £32,342 million, or 13.8 per cent of total consumer expenditure (Heasman 1989: 12). In 1960, total food expenditure (at current prices) was £4,485 million, or 29.1 per cent of total consumer expenditure. At that time, processed food was estimated to account for 68 per cent of household food expenditure. Today, processed food accounts for an even larger share of food expenditure: only eggs, meat, fresh fruit and vegetables are significant food items which do not require processing. The potential offered by sophisticated food processing has still not been achieved. A product like sugar, which might be expected to be adversely affected by rising incomes, has been transformed into so many products, that its presence is barely recognizable (Heasman 1989: 9). Although per capita demand for sugar has risen most quickly in the developing world, in the industrialized countries, too, sugar has retained an important position in the diet. Agribusiness has elaborated new 'standards' of food consumption, as we shall see in Chapter 4, and popular diet has been reorientated towards foods which require a heavy processing component.

Increasing product differentiation has been allied to continuing concentration among food companies. This oligopolistic control of the food market is most pronounced in food products with the least nutritional value, such as chewing gum, sweets, biscuits, cereals, cake mixes and carbonated soft drinks, all of which, of course, require sugar as a major component. Food products can be developed, then, to counteract, at least in part, changes in modern diets, or to develop new tastes for foods which call on traditional ingredients like sugar.

International comparisons of current food consumption also suggest a growing similarity in dietary patterns (Frank and Wheelock 1988: 24). In most developed countries the consumption of dairy products, meat, eggs and sugar has increased, at least initially. There is also an increase in the proportion of fat and simple sugars in the diet. At the same time the consumption of potatoes and cereals has decreased; a marked decline in the proportion of dietary fibre has been observed in all developed countries.

In the United States changes in food consumption have followed changes in the labour market. As Belasco observes, the decline of the 'family wage' concept, as more households consist of dual-income earners, was reflected in the development of new 'convenience' foods. Convenience foods 'enabled women to snuggle up to their husbands on the couch', rather than slave in the kitchen, at least in the perception of advertising and marketing men (Belasco 1989: 156-7). In most of Europe, too, married women now constitute a significant proportion of the total labour force. As we have seen, families with two adults working outside the home have less time to dedicate to preparing meals, but they also possess the means to acquiring relatively expensive animal products, high-value processed foods and they tend to eat out more frequently. If we look at changes in food consumption in the United States it gives us some indication of the way diets in Europe are likely to change in the next decade or so, and many of the changes are already pronounced. In the period from 1960–3 to 1980–3 high-value products showed significantly increased consumption in the United States: frozen potatoes, broccoli, corn and orange juice all registered increases of between 118 per cent and 420 per cent. Some fresh products (avocados, pineapples, nectarines) were also major beneficiaries of changes in diet. Canned food consumption declined in the same period – suggesting that canning represents an earlier 'stage' in the evolution of the processing industries. Most interestingly, meat consumption dropped dramatically, especially lamb and veal, although chicken consumption increased by 76 per cent in the twenty-year period (Frank and Wheelock 1988: 27). In the United States, and now in Europe, a profound shift is taking place towards eating 'healthier' foods, even if they are foods in which value added is a significant ingredient (Daly and Beharrell 1988). 'Value added' refers to the 'value' that is added to food during processing and distribution. Even 'healthy' foods are often heavily packaged and, sometimes, processed, too. This trend has implications for marketing strategies within the modern food system, but also for other behavioural changes, at the individual and household levels, such as patterns of work, recreation and public attitudes to what are perceived by some as Green issues. These are discussed more fully in the final chapter.

In Britain the changes in food consumption have been no less pronounced, and broadly reflect those for the industrialized countries as a whole. For the period from 1951 to 1983 the average real weekly earnings, the purchasing power, of adult male manual workers almost doubled (Burnett 1989: 301). Among other things, this brought an increase in home ownership, and expenditure on the home increased dramatically. Kitchens have been enlarged, as we have seen, and the rise in consumer durable goods ownership has been maintained for the majority (although by no means all) of the population. In 1986, nine out of ten households had a television and a refrigerator, eight out of ten a washing-machine, and seven out of ten had central heating. 'Home-centredness' has become almost an article of faith in advertising directed towards filling more and more private space with more consumer durables, and the products (especially foods) which can be used together with them. The wider community aspects of this 'privatization' of assets and social relations is discussed more fully in the next chapter.

Not surprisingly, major shifts in food consumption reflect changes in household composition, and British social structure exhibits some interesting features in the late 1980s. What was once considered the typical 'breakfast cereal' family, consisting of a husband, wife and two children (usually one of each sex!) now accounts for only 11.9 per cent of households. At the same time, one-person households now account for 24 per cent, and two-person households for 30 per cent of total households. Large families (that is with four or more children) have almost disappeared (1.9 per cent). With changes in family composition eating habits have changed, and food consumption now reflects specialized tastes and needs, or the catering needs of small groups. It is no accident that the microwave oven, ideal equipment for cooking dishes for one or two people or the ubiquitous 'television snack', has become so important a feature of so many homes in such a short time. The microwave is particularly important for shift workers: indeed, it might be argued that the microwave brought the eating habits of shift workers into line with those working conventional hours. In terms of food expenditure, household size is now a more important factor even than social class. In 1985 the poorest 20 per cent of the British population spent an average £7.49 per capita on food, 29.4 per cent of its budget. By contrast, the wealthiest 20 per cent spent an average £9.87 per head on food, 15.5 per cent of total personal expenditure. The differences in food consumption are much more marked if we look at the size of the household. In 1985, households consisting of single adults, living alone, spent more than twice what each member of 'family' households spent on food (Burnett 1989: 303).

As we have seen, shifts in British food consumption fit squarely into those for developed countries as a whole, but do deserve some comment. Foods which have increased in consumption in recent years – pork and poultry, wholemeal bread, margarine, pasta, breakfast cereals and frozen foods – 'do not fit into a single category or explanation' (Burnett 1989: 308). Health consciousness is certainly a factor in the rise in importance of some of these foods, but convenience also assumes importance (the consumption of fish declined by 21 per cent between 1962 and 1976, despite the emergence of 'fish fingers', the best that packaging could do!). In addition, it is clear that popular diet has been transformed by more exposure to foreign foods: rice and (especially) pasta have leaped into the British diet in a way unimagined even twenty years ago. (The increase in the consumption of pasta was 160 per cent between 1962 and 1976). British people have also been exposed to the foods made and prepared by immigrants, such as Indian curries, which have so enriched what many see as the traditionally bland British diet. It is important to note however, that, consistent with the changes in women's labour referred to above, the single most important aspect of change in the British food consumption has been the rise of convenience foods. Between 1980 and 1985 increased expenditure on convenience foods was almost double that for food as a whole (Burnett 1989: 309).

Changes in food consumption are, in turn, connected with changes in food retailing, an area like food consumption, in which convenience has played a major part in shaping current practices. The earlier description of a 1930s housewife shopping on a daily basis for food supplies, with which this chapter began, bears little resemblance to food shopping habits today. Changing attitudes towards the use of time have led to an emphasis on convenient shopping, as much as convenience foods. People today make fewer shopping trips and visit fewer locations to shop. Increasing car ownership has made the shopping population more mobile and has increased the demand for 'one-stop shopping'. Higher income consumers, who are geographically mobile, are prepared to travel some distance to their preferred retail outlets, while less mobile, poorer consumers are confined to shopping locally, where food prices are relatively high. The food retailers have responded, and partially created these trends, by creating larger but fewer retailing outlets to cater for the increased number of product lines, and the simultaneous demand for convenient shopping. The market share of large supermarkets and hypermarkets has, not surprisingly, grown in recent years: in 1988 they accounted for about one quarter of food purchases in France, West Germany and the United Kingdom (Dawson *et al*. 1988: 51). Second, all the major retail chains in the United Kingdom (or 'multiples' as they are called in the jargon) sell a significant proportion of 'own brands' (as high as 30 per cent of market share of groceries).

Table 1.6 UK grocery retail market shares 1987–8

Retailer	Market share (%)
Sainsbury	13.7
Tesco	12.9
Gateway	11.3
Argyll	9.5
Asda	7.4
Other multiples	23.2
Co-op	12.9
Symbols	3.6
Independents	5.5
Total	100.0

Source: Duke (1989) from AGB/Mintel

Table 1.7 Historical margins in grocery retailing

Year	Margin: all multiple food retailers %
73–4	3.5
74–5	2.5
75–6	1.7
76–7	2.1
77–8	1.8
78–9	1.7
79–80	1.9
80–1	1.8
81–2	2.2
82–3	2.2

Source: Duke (1989) from Institute of Grocery Distribution

However none of these multiples (see Table 1.6) is pursuing 'backward linkages' as a way of integrating their retail activities with food processing. Instead food retailers have diversified into other markets, such as clothes and 'do-it-yourself', which have enabled them to escape from the relatively static (in terms of turnover) food market, and to pass on the costs of product development to the processing industry (Carter 1989). Their commercial objectives, to develop 'value added', to increase very tight margins (see Table 1.7), and to expand the market in money terms, can best be sought through increasing product (brand) identification. Currently the multiples are much more interested in targeting each other's customers than in pioneering changes in processing food (Duke 1989). The degree of concentration that exists in food retailing (see Table 1.8) – the six large multiples, including the Co-op, account for 70 per cent of the market – and the low margins in food retailing provide the backdrop for extensive product differentiation and, increasingly, the accent on health, convenience and 'difference'. At the same time, stores such as Marks and Spencer, with a high 'quality' profile among consumers, have been able to move from the sale of clothes to relatively expensive foods, carefully packaged and appealing to the upper end of the market. These trends are explored in the concluding chapter, as evidence of changes in consumer consciousness, linked,

Table 1.8 Percentage share of total grocers – long term – UK

	Multiples	Co-op	Symbols	Other independents
	%	%	%	%
1976	57.1	20.4	12.4	10.1
1977	60.5	18.9	11.4	9.2
1978	64.1	17.8	10.0	8.1
1979	65.9	17.6	9.2	7.3
1980	67.5	17.6	7.9	7.0
1981	69.6	16.9	7.2	6.3
1982	71.9	15.8	6.3	6.1
1983	73.5	15.1	5.7	5.7
1984	74.9	14.2	5.0	5.9
1985	N/A	N/A	N/A	N/A
1986	77.1	13.5	3.9	5.5
1987	78.0	12.9	3.6	5.5

Source: Duke (1989) from AGB

among other things, to the progressive 'Greening' of issues like food and the countryside.

This section has examined the long-term trends in food consumption in Britain against the background of other industrialized countries. We have indicated that food products, and with them eating habits, reflect rapidly changing social patterns: smaller households, women in waged employment and less time allocated to food preparation and shopping. These trends, and others, have assisted the full development of the food system, by enabling women to find work (especially in the food processing industries) and by allowing families to increase, and diversify, their pattern of consumer expenditure, providing a rapidly enlarged market for white goods, and other expensive durable goods. As we shall see in Chapter 3, the move from extensive to intensive accumulation within modern capitalist society has depended on two factors above all else: a freeing of disposable income and greater integration between the production and consumption 'ends' of the food system. This integration does not necessarily imply wider ownership of capital, but it does imply a series of production incentives, and consumer promotion strategies geared to increasing or defending profitability. Some of the difficulties that have accompanied these structural, market-based transformations, notably their impact on the natural environment, are discussed in Chapter 5.

CONCLUSION: THE TRANSFER OF LABOUR AND THE PRODUCTION OF FOOD

This chapter has argued that changes in the preparation of food, outside and inside the home, need to be related to shifts in the labour process, which have taken women into more wage employment and brought new domestic technology to bear on housework. More and more household tasks – the tasks of servants in nineteenth-century middle-class households – have been handed over to outside specialists. This process, which reached deep into the social fabric of British and other industrialized societies, mirrors the wider process through which industry has appropriated the domestic labour process. In being commoditized, domestic labour was converted into an arena for accumulation. The development of new food commodities and white goods technology, has shifted attention (for the household as well as capital) from production to consumption. The development of new patterns of leisure and work, together with new patterns of food consumption, have proved crucial for the full development of the food system. These social patterns have brought with them important ideological dimensions, a point to which we shall return in Chapter 7. They should also lead us to reflect on the fact that the developing food system cannot be defined simply in terms of technological and economic restructuring. Ideology is a potent force within the food system today, as we have seen, but the ideological component of changes in diet and the domestic division of labour has received little attention from sociologists. One way of approaching the issue is to compare the competing ideologies associated, on the one hand, with women's 'independence' and, on the other, with the role women play in providing the nurture and support for other members of the family – the 'career woman', whose influence as a consumer is valued by advertisers, and Katy, the original Oxo housewife, whose principal contribution to the family consisted of providing an appetizing meal for her husband and children.

The central thrust of our argument, to which we will return, is that a correspondence occurred, as the modern food system developed, between changes in the labour process and changes in food preparation (and housework generally) in which women's roles must be seen as pivotal. Without more women in paid employment outside the home the post-Second World War 'consumer revolution' could not have occurred; and one of the principal elements in this 'revolution' has been changes in the way housework is performed. Women's labour has enabled domestic technology to have a longer reach, and with this longer reach technology has had a transformative power, not perhaps to reduce the *overall* burden of women's work, but to change its nature and its effects. As labour has been transferred from the home and become a commodity, food has been commoditized with it, and the transformation of use values into exchange values has come about through major shifts at the ideological level, as well as in economic terms. It is difficult to exaggerate the importance of this question; today our diets are radically different from those of a hundred or even fifty years ago. The majority of women routinely use food that has been processed and prepared outside the home, in conjunction with an ever-increasing number of devices (from refrigerators to dishwashers to microwaves) which have transformed the nature of housework. Outside the home, the demand for fast food and restaurant meals has escalated without parallel, making the food industry the largest in the economy, and dwarfing the agricultural production sector which originally defined the food system along more local-ized, 'natural' lines.

This chapter has examined the 'naturalization' of women's labour, and its relationship with the commoditization of food. In the next chapter we shall examine the contribution of farming to the development of the modern food system. In doing so we shall again confront the question of women's labour within the family, and the relationship between increased commodity production and the specialization of women's work. For the moment it is important to register that the food system developed around structurally compatible processes, changes in technology and changes in the labour market, and that these processes were neither confined to the home nor the factory, but spanned both. The redeployment of labour – and technology – as the food system developed, was not the result of technological imperatives alone, nor was it entirely socially determined. It was the result of a con-vergence of several processes and several components whose central feature has been the transformation of women's labour, as the means, the vehicle, through which radical changes in food production and preparation could be effected.

The next chapter takes up the argument from the point of original production, the farm household rather than the urban consumer. We have seen that many of the tasks traditionally performed by women within the urban household have become more specialized and differentiated within the modern food industry. The picture that emerges in the following chapter is a rather different process; the farm family has survived by using the domestic unit, the household, to accom-modate to external specialization (in crops, technology and capital). Farms have become more specialized, and the part played by food and fibre production within the wider food system is much narrower than in the 1930s, but family labour

45

on the farm, particularly that of women, has enabled farmers to shed salaried labour and yet make closer connections with the industrial agri-food suppliers and marketing chains. Again, 'naturalization' can be invoked, since the 'family farm' has acquired greater legitimacy while its structure was undergoing radical change. The only serious threat to this legitimacy is currently being posed by the damage modern farming practices do to the environment and, as we shall see in Chapter 6, the rising consciousness of environmental issues is attributable, among other things, to the loss of sustainability on the farm. Again, the components of the system need to be viewed as a whole if their wider significance is to be appreciated.

2

THE PASSING OF RURAL SOCIETY

We have seen that changes in food consumption, and employment for women within the food system, can be understood in terms of the convergence of technological change, with social and political forces, inside and outside the home. Underlying structural processes, through which food became a commodity, an industrially produced good, were linked to a transfer of women's labour from housework (and domestic service) to the wage-labour market. This transference was facilitated by the development and adoption of new 'domestic' technologies which have materially affected the content (if not the quantity) of housework, including food preparation in the home. To appreciate the full impact of mass consumption we need to consider the roles of women within the family and the wider economy. This requires an appreciation of the changing role of women in the light of new food technology, especially in the food processing, marketing and retailing sectors. For the moment our attention is directed to the point of production: the farm economy and its links with the encompassing rural society.

This chapter examines the effects of greater integration in the food system at the point of food production: the farm. We explore the 'production end' of the food system by concentrating attention on the way 'family farming' has developed, with the appearance of stability and coherence, within the rapidly changing English countryside this century. The conditions for change in farming were established by the demographic transition through which England passed during the nineteenth century, which effectively transformed a rural into an urban society. In this chapter we examine these

47

changes, initially in terms of their demography, but also in terms of the ideological convictions and issues to which urbanization gave rise. Following the move from the land, British agriculture entered a prolonged crisis, when land was neglected and no easy solutions to the farm crisis were evident. At the same time, and partly because of depressed prices and earnings, freehold ownership of farms increased. In turn, partly as a consequence of this trend in owner-occupation and of agriculture's strategic importance in wartime, it came to occupy a privileged position within the economy. Since the last war the agricultural sector has experienced a decline in numbers, but also a revival of confidence; as a result it has come to be associated with continuity in post-Second World War society.

An examination of the labour process on family farms takes us into a discussion of the capacity of family holdings to accomodate to technological change, enabling farmers as a group to acquire greater political weight, but at the same time leading to important changes in agricultural husbandry. To avoid being 'marginalized', family holdings had to be increasingly competitive. Beginning with an account of 'mixed husbandry' in the 1930s, the chapter considers recent developments in relations between the farm and the market, which have contributed to the breakdown of 'natural' organic cycles, and the loss of traditional mixed farming, a system which combined food production with environmental protection.

At the same time, through the increased dependence of farmers on external suppliers, economic linkages have been developed with product and input markets which serve to weaken the entrepreneurial control of the farmer. The transformation of family farming and the acquisition of freehold ownership have also been the means through which capital has entered the agricultural sector, without leaving most producers landless. This process of capitalization is, in turn, linked to the role of the state in regulating the terms under which agricultural production takes place, particularly the markets for farm products (an issue taken up in more depth in Chapters 3 and 6).

This chapter is primarily concerned with the farm household and the social milieux that have accompanied these changes in agriculture. The social structure of rural areas has altered fundamentally since modern agriculture became established after 1945. Today, the agricultural problem is in a minority, but the 'countryside' is a pole of attraction for many urban people. For most of them it is essentially an arena of consumption rather than production. At the same time, behind the resonance of the English countryside lies a reality of rural social deprivation, a picture which has been obscured, just as the vision of the countryside has been distorted, to reflect the ascendancy of new class interests.

Environmental concerns have also become one of the ingredients of the new countryside, but, as we shall see in Chapter 6, farming as an activity is seldom practised in an environmentally sustainable way. As the sustainable dimension is lost to agriculture, the locus of political dissent moves from capital and labour on the farm to the urban consumer of food products and environmental goods, from 'nature' and rural space to industry and the urban nexus. The implications of these changes are taken up later. Here the focus is family farming and rural society, both of which have claimed greater ideological importance at precisely the historical moment which marks their transformation.

THE RURAL LABOUR PROCESS
AND THE FAMILY FARM

The rural labour force: a demographic transition

As John Saville (1957) pointed out in his classic study of rural depopulation in England, the discussion of social changes in rural England is dogged by confusions surrounding the term 'rural'. The population caught up in the process of urbanization during the nineteenth and twentieth centuries left behind them a rural society, but 'an accurate definition of what constitutes a rural district is not easy, and the difficulty of definition increases as we approach our own times' (Saville 1957: 3). One approach, that taken by Saville, is to follow official designation, as incorporated in local and national government legislation. Saville traces the administrative distinction between 'rural' and 'urban' to the Public Health Act of 1872, which made the size of settlements a key factor in local government. This provided the framework for the

Consolidating Act of 1875 and the Local Government Act of 1894. The difficulty, even at the end of the nineteenth century, was that urban overspill had transformed rural areas into urban ones, without necessarily changing their designation. By the 1920s and 1930s this urban overspill was already evident in an 'increased' (in official terms) rural population. Other attempts to distinguish 'rural' from 'urban' areas – in terms of land use, occupational structure or social networks – all present analytical problems every bit as intractable as settlement size. As we shall see later in this chapter, they also account for the difficulty in defining and assessing social deprivation in rural areas today.

It is important to establish at the outset that urbanization is not synonymous with *factory production*. It was not until the late nineteenth century that factory production really became dominant in Britain, but urbanization had already changed the face of the country. Most production in urban areas was small scale, and often based on simple commodities, using the family, in an extended form, as the model for the enterprise. In the next chapter we emphasize that 'agriculture' is not a 'natural' process, but one which is established under historical conditions, incorporating elements of 'nature' together with elements of industrial processes. It is worth indicating at this point that industry, no less than agriculture, is a historical form, with contours and characterisitcs not easily reduced to nineteenth-century factory production.

Rural–urban migration began to assume importance in England after about 1780, when the population in urban areas began to rise, in some cases dramatically. However, not until the 1830s and 1840s did these migratory trends begin to affect the absolute population of rural communities. After 1850 a sizeable number of rural parishes began to experience an almost continuous decline in their total population, but it was not until the last two decades of the nineteenth century that urban population growth was unequivocally linked to declining population in most rural areas. At this stage the view often held was that the 'agricultural depression' accounted for the abandonment of rural areas, although the truth was rather different. It is true that the natural increase in urban population was reflected in urban population growth before most rural areas experienced decline, but the growth of urban industrial employment had sent shock waves through rural society before total population in rural areas began to diminish.

Rural depopulation rose after the 1840s because of a combination of factors. Employment opportunities in the countryside were more difficult to find as the nineteenth century progressed. Many of the economic activities traditionally performed in villages had been transferred to the towns, and larger economic units came to replace smaller decentralized units, which combined 'industrial' and 'non-industrial' (agricultural, domestic) functions. This process of structural relocation and development was reinforced by administrative and government arrangements. By the late nineteenth century, the greatly enhanced powers of national and local Government had been located in towns.

What this implied for the countryside was that agriculture, which had always been the principal economic activity, now became almost the only activity, at least in many rural areas. Already, by 1820, enclosure had transformed the rural map of most of England, having received its final stimulus from the Napoleonic Wars and the quest for increased food. There was little public use of common land, and few resources held 'in common'. The three-tier system of landlord, tenant farmer and labourer (which is discussed in more detail in the next section) was widespread, if not universal, throughout England. By 1850 the existence of a surplus rural population, which could not be absorbed by agriculture, began to be reflected in rural parishes, as we have seen. The potential mobility of this surplus rural population was also important, because the development of a railway network had opened up the prospect of maintaining contact between urban and rural areas for the first time.

It is important to remember that rural labourers had traditionally been subjected to certain restrictions on their freedom to move. Since the seventeenth century the English Poor Laws had determined that rural labourers could only receive poor law relief in their place of residence. These restrictions, the Settlement Laws, were already interpreted much more flexibly by the late eighteenth century, but they were not undermined, finally, until the rate of urbanization gathered momentum half a century later. Restrictions on the freedom of movement of labourers and their families underlined a principle, even if it was frequently not practised: people 'belonged' to their parish, and the parish assumed some responsibility for their welfare. It is worth mentioning that as late as 1907 about 12,000 people were being removed annually from London and other large cities and returned to their rural parishes.

It is also important to underline that the rural exodus was associated with changes in farming, as well as urban industrial employment. It was 'push' as well as 'pull' that led people to leave rural areas. As English agriculture broke into its stride, notably in the period of 'High Farming' between the 1850s and 1875, farming as an activity required a different kind of labour force. During the 1850s and 1860s much English farming gradually converted from cereals to dairy and beef production, partly under the influence of North American competition in cereal production. From the middle of the century onwards farming became a more intensive activity, more mechanized, using more inputs (some of them not available on the farm) and more sensitive to the market. These changes brought benefits to most farmers; only agricultural labourers languished in the shadows of this transformation. Between 1861 and 1881 the farm labourer population declined by 20 per cent. Significantly, the number of women and children working in agriculture dropped most steeply. As male labour was substituted for women and children, the increasing costs of labour persuaded farmers to substitute capital for labour, where they could.

This substitution of labour by machinery increased the 'push' factor in migration. The wage differential between agriculture and industry, which had always been important, began to assume critical importance for a predominantly male wage labour force, shorn of its petty-commodity production character. The 'crude wage ratio' between agriculture and industry was about 50 per cent for most of the second half of the nineteenth century. Between 1881 and 1901 male agricultural labour declined further, by over 40 per cent. It increased slightly after 1901, but never regained its earlier strength. By the early twentieth century there were far fewer farm labourers, most of them were men, and they were clustered in the minority of 'agricultural' counties. Female labourers had, in the main, 'been withdrawn by their husbands and parents from field work', as Lord Eversley commented in 1907. In 1851 there had been 143,000 female farm labourers; by 1901 there were only 12,000 left. With this decline in the female agricultural population, which moved into towns to swell the ranks of domestic service, went a decline in 'female' trades, such as lace-making, straw-plaiting and the making of clothes. As we saw in the last chapter, the transfer of women's labour from the home to outside employment was an essential feature of the modern food system's expansion. In an earlier phase, it should be remembered, many women left farming itself. Women left wage employment in agriculture, but this did not necessarily mean they left rural society. As we shall see, women remained central to the functioning of the farm family and the farming economy.

The family farm

Family farming dominated British agriculture in the years between the First and Second World Wars, but in many respects the family farms of the 1930s were radically different from those of today. Agriculture was depressed, prices were low and few of the tenant farmers who still characterized British farming saw their farms as profitable businesses.

British agriculture had experienced important changes during the previous century. Farmers were more market-orientated, producing for the rapidly increasing urban population. They practised husbandry techniques of which their forefathers were unaware, they used chemical fertilizers to a limited degree and, increasingly, they used machinery for draught power. They no longer met their own food needs from their own production. They had benefited from 'convertible husbandry', growing crops partly to feed animals, and grazing animals on pasture. The family farmers of the 1920s and 1930s produced milk and meat products which could be transported by rail and refrigerated. They were part of a depressed sector, but one which had already come to terms with the market. René Dumont presents a vivid picture of the English countryside during the inter-war years, the era of 'stick and dog' farming when the farmer managed his holding relatively extensively 'almost content to walk his land all the year round, inspecting his herd, which his well-trained dog paraded before him' (Dumont 1970: 379). The view at the time was that the countryside was laid waste by the Depression, with a marked visible impact, especially in marginal areas. Walls and hedgerows prevented animals from straying and were maintained mainly for that reason. However, carefully tended and manured grassland was a rarity, thorns and thistles were left uncut and much of the arable land reverted to rough grazing, as bracken spread its hand over the depressed landscape. The prevailing view was that without a dynamic agriculture the countryside would be transformed into wilderness. The implicit assumption was that a 'farmed' or 'managed' landscape was the only viable alternative. The revival of the countryside, in the prevailing view, waited on the revival of farming. The contrasts with the situation soon after the Second World War are instructive. The post-war 'settlement' at that time assured British agriculture a security denied to it since the Corn Laws were repealed a century before. What followed, as agriculture geared itself up to meet new production possibilities, represented a completely new kind of threat to the countryside, and one that was not well understood until the agricultural industry was faced with politically embarrassing surpluses.

The post-war settlement, after 1945, lifted farming out of the pre-war doldrums, and consolidated war-time 'prosperity'. In 1931 there had been over 700,000 full-time farm workers; by 1956 there were only half a million, the only major industrialized country with as little as 5 per cent of its working population in agriculture (Newby 1977: 81). By 1953 the number of draught horses had fallen to less than half the pre-war figure, and farmers bought fuel for their tractors at a special price of a shilling a gallon. Taxation was waived on agricultural intensification, which was also encouraged by shortages of labour. Between 1937 and 1951 the cost of machinery, as a proportion of total costs, increased from 6 per cent to 16 per cent. The costs of labour, again as a proportion of total costs, increased from 27 per cent to 41 per cent, this forced farmers to make greater economies in their use of labour (Saville 1957: 142). Dissecting the farm accounts of some Shropshire farmers, Dumont noted that the sale of milk brought in £1,620 on one farm in 1950, as against £700 in 1945. He comments, with what seems like innocence, in view of future trends, that 'an association of producers, the Milk Marketing Board, regulates prices, but it operates under the control of the government which watches the interests of the consumers' (Dumont 1970: 386).

The restructuring process had preceded state assistance via a long-drawn-out cost–price squeeze. Farming had already ceased to be the country's major industry in 1911, when only 7.6 per cent of the working population remained in agriculture. But it was not until forty years later that farmers were able to depend on government to guarantee prices, markets and credit at low rates of interest. Indeed, the transformation of tenants into freeholders was still underway in 1951, twenty-five years after one prominent industrialist had argued for a shift towards owner-occupation:

> I would not interfere with the freehold ownership of the farmer today, or the possibility of the farmer tenant today becoming the freeholder when he wanted to. He ought to have that option, and state credit to enable him to buy his land on reasonable terms.
>
> (Mond 1927: 295)

During the 1920s and 1930s the principal political division was between those, especially Liberals, who favoured local authorities buying up land and leasing it to farmers, and those who favoured sitting tenants being able to buy up their farms.

What Arthur Young had called 'the magic of property' did not touch most English tenant farmers until after the Depression, but the notion of a 'property-owning democracy', so central to Tory thinking then as now, was to become a central element in the transformation of English family farming. After the Second World War the gradual shift towards freehold ownership of farms helps explain why farmers embraced the modern food system.

Something of the flavour of English agriculture during the inter-war years can be gauged from a study of farming in one Yorkshire village, Hooten Pagnell, published in 1934 by Ruston and Witney, two pioneering agricultural economists. Hooten Pagnell was unusual in that all the farms were tenanted, and located in the village itself and a neighbouring hamlet, but in most other respects the face of farming in England in this period is vividly revealed in Ruston and Witney's work. There are five aspects of family farming in 1931, when the field research was undertaken, which establish the essential characteristics of agriculture agriculture at this time, and help us to measure the impact that the industrial food system was to have in transforming agricultural priorities:

1 the system of land husbandry, which still followed organic cycles and clearly demonstrated the limitations presented by biological time and space;
2 the relatively 'extensive' system of production having only weak links with outside markets, and poorly developed 'backward' and 'forward' links with industrial suppliers and processors;
3 the dominance of tenancy over ownership, and the implications which the absence of 'freehold' acquisitions held for farm modernization;
4 the family labour process, and the continued dependence on family labour in the farm enterprise;
5 the limited degree of state interference in the regulation of markets for farm products.

In Hooten Pagnell, we can also identify some of the fundamental strengths which enabled family farming to survive a period of economic crisis.

Hooten Pagnell (1931)

The dozen farms that accounted for Hooten Pagnell's agricultural population were characterized by mixed husbandry. In the past farming had been less well equipped to take advantage of arable crops for animal feed and relatively productive grassland for grazing. Farms had been created out of 'unenclosed' strips, a process which had not been fully completed until 1917, when the strip remnants of intermixed ownership had finally disappeared. With more contiguous holdings, a new 'freedom to farm' had developed, although farmers remained tenants of an ancestral landowner. This entrepreneurial freedom was highly prized:

> If there is one thing which differentiates the modern farmer from his forerunners, it is his freedom – freedom to crop the ground as he wills, freedom to market his produce when, where and as he wills, freedom to devote his time and money and energies to furthering his own interests as he thinks fit, although signs are not not wanting that 'this freedom' may not last for ever
>
> (Ruston and Witney 1934: 15–16)

Tenants' freedom to practice land husbandry as they wished had been a consequence of the Agricultural Holdings Acts, and had provided farmers with more entrepreneurial control over their enterprises than ever before.

Output was low on Hooten Pagnell's farms in 1931, although not much lower than in the country as a whole. Land was not worked intensively. For each hundred acres of farmed land only £136 was spent on 'inputs' (feedstock, artificial manures and seeds) compared with an average £180 for farms in England and Wales. This was not much more than a fifth of what was received from marketed production (£625), a proportion similar to that for England and Wales as a whole. Half a century later farmers in the country as a whole spent more than a third of what they received from farm sales on 'inputs' of this kind (Nix 1984: 153).

In Hooten Pagnell, in 1931, farms were dependent for most of their income on livestock and dairy production, but stocking levels were low. Just over half the cultivated land was ploughed and the rest was left to permanent grass, but of the land under the plough most was either dedicated to low-price cereals with low returns, to feed crops for livestock consumption or to 'cleaning' crops like turnips. As in England and Wales generally, the stock-carrying capacity of land was low; farmers bought few inputs and used little labour, the land supported low stock ratios and few high-price crops were grown for sale. Farming was 'mixed husbandry', there was little specialization and no incentive to intensification.

Within this overall pattern, of course, there were important variations. Even within one activity like dairying, individual farmers followed their own inclinations:

> One rears his heifer calves with the object of eventually drafting them into the herd and cutting down depreciation. Some retain their own milk, in the village or neighbouring town; some wholesale it, possibly at a distance . . . some depend almost entirely upon the home production of foodstuffs for their cows, and others almost entirely upon purchased concentrates and grass . . . but no two milk producers manage their herd in exactly the same way.
>
> (Ruston and Witney 1934: 38)

Prices and markets were depressed, and on average only a third of agricultural output was marketed. Farming was an activity which, although less 'self-sufficient' than in the past, was still determined by the farmer's preferences, with little involvement from outside advisers or technical assistance.

The depressed state of agriculture was reflected in rents, which were stable in 1931, but low, only a little above pre-1914 levels. However, as Ruston and Witney make clear, the landlords' returns were much less than in the pre-war era, because taxation took more. In 1931 the landlord retained barely a third of what his tenant paid him; in 1913 it had been over half. Partly as a result, the shift away from farm tenancy was underway in England and Wales. In 1913 over 90 per cent of farmers were tenants, by 1927 it was 64 per cent. In the pre-First World War period, agriculture in the United Kingdom exhibited a relatively high proportion of tenants to owner-occupiers. At this time under a half of French farmers were tenants, and little over a quarter of farmers in the United States. The United Kingdom was beginning, slowly, to fall into line with other countries.

During the so-called 'golden age' of English farming, in the mid-nineteenth century, agriculture had depended heavily on landlords for capitalization and more rational production, but, after the First World War, falling land values, increased taxation and lower product prices removed the landlords' ability, and willingness, to equip and maintain the land. The 'tripartite system' of landlord, tenant farmer and labourer, referred to by Marx in *Capital*, was gradually disintegrating:

> The landlord, where he still remains, is in many cases financially incapable of the replacement and maintenance of the permanent equipment at a proper level of efficiency. The occupier, where he is also landlord, is even less capable in this essential matter . . . He is finding, moreover, that possibly the worst and least sympathetic of landlords is 'borrowed money'. As a tenant in bad times the landlord met him in a rent reduction; as an owner with his property mortgaged, mortgage interest still has to be met, good times and bad, under threat of foreclosure.
>
> (Ruston and Witney 1934: 229–30)

Landowners were experiencing difficulties in continuing to pay death duties, in addition to increased rates and taxes. Tenants' rights were protected, but there was little incentive for modernization. Not until the Second World War was English agriculture to feel the necessary stimulus to increase production, and not until the post-war period were many farmers to enjoy the privileges and risks that accompanied owner-occupation.

The family farm as constituted in Hooten Pagnell, had a number of important features. It was family labour which remained the basis of the farm enterprise, and the farm which served to reproduce the family. The 1931 Census revealed that there were forty-one labourers to thirteen farmers in Hooten Pagnell, compared with sixteen farmers and seventy farm labourers in 1851. In this eighty-year period the agricultural labour force had contracted considerably: 52 per cent of the 'economically active' population were employed in agriculture in 1851, but by 1931 it was 37 per cent. There had also been a small decline in the number of rural tradesmen, whose work was closely associated with farming, in the same period.

A close look at these figures, however, reveals that family labour played a much larger role in determining the conditions of existence of family farms than might at first be imagined. Of the forty-one farm labourers in Hooten Pagnell in 1931 only twenty-four were regularly employed, the others were casual labourers only. This gives a ratio of 1.43 workers to every 100 acres of farmland in Hooten Pagnell, roughly half the average for England and Wales in 1931. Furthermore, of these twenty-four regular farm labourers ten were relatives of the farmers for whom they worked, often living with the farm family in the same house. What we see in Hooten Pagnell is a farm labour process constructed around the nuclear farm family, with limited additions of waged labour, mostly employed on a seasonal or casual basis. The implications of women's labour for the process through which family farms have become capitalized is discussed in greater detail in a later section.

LAND OWNERSHIP AND
POLITICAL LEGITIMACY

By the 1880s, grain from North America was flooding into the United Kingdom, persuading most countries in Western

Europe to adopt protectionist measures, although Britain was relatively late in joining them. Wheat farmers, especially in East Anglia, left more of their land in permanent pasture. In the meat and dairy sectors the impact of North American competition was felt less strongly, and the prices received by producers were often maintained. As we shall argue in the next chapter, the enlargement of world agriculture which was signalled by the development of the Great Plains cereal production, was to produce a major transformation in the food system, making it, effectively, a global food system and one which eventually would lead to new patterns of food production for a mass market, as well as changes in the labour process.

The process of restructuring within British agriculture was momentarily reversed during the First World War, when food self-sufficiency was again a policy goal of a besieged government. After the First World War as we have seen agricultural decline continued. The question we need to address now concerns one of the other major factors in the subsequent restructuring process: what effect did changes within agriculture, and particularly increased owner-occupation, have on the legitimacy of the sector as a whole? To what extent have changes in the food system since the Second World War been linked to changes in the political credentials of farmers and landowners, within the broader society? The answers to these questions lie, to some extent, in an analysis of changes in land tenure during the previous century. It also helps explain the link between greater economic efficiency and the reduced sustainability of British agriculture.

The ownership and concentration of land need to be looked at in conjunction. The ownership of land has always conferred great advantages, in pre-industrial England as well as in the eighteenth and nineteenth centuries, but it was not until the Industrial Revolution that landowners were offered the opportunity of establishing themselves as fully-fledged capitalist entrepreneurs. The legitimacy of landowning in Britain does not derive simply from the fact that so many farmers today are owner-occupiers; it also reflects the remarkable flexibility and continuity of the landed estate (Anderson 1987).

As we have seen, by the mid-nineteenth century farmers and agricultural workers were in a minority among the 'economically active' population, although agriculture was to remain the single most important source of employment until much later. Only a fifth of the working population was employed in agriculture in the 1850s. The capitalist mode of production was already well established in industry, and agriculture was, to a significant extent, the beneficiary of this process. Anderson reminds us that, by the mid-nineteenth century, farming was relatively productive by continental standards, and financed partly out of urban land speculation and activities such as coal-mining:

> Britain was the largest coal producer in Europe throughout the nineteenth century, and its mines fell predominantly within the estates of major agrarian proprietors, capable of initial investments in them, and then of securing advantageous leasing arrangements to industrial venture capital. Most significantly of all, the English aristocracy played a role in urban development which had no equivalent anywhere on the continent.
>
> (Anderson 1987: 32)

In some parts of England, such as Norfolk, there had been a considerable integration between agriculture and related industries for some time. On some estates, mechanization was already advanced by the 1850s, and in the subsequent decade the introduction of the mechanical reaper, and then the reaper-binder, reduced the size of the labour force needed for harvesting (Martins 1988). By the 1850s in 'advanced' areas like these, agricultural science was also making a considerable impact. The need for the four-course system was challenged as new chemical fertilizers made it possible to grow crops in succession, without reducing the fertility of the soil. By the 1860s some farmers were already seeking – and securing – permission to depart from the terms of their leases, to grow more grain. (Jefferies 1987: 68–9).

Livestock production became more intensive as the value of livestock products rose. In the more 'progressive' areas of East Anglia, farmers increasingly used imported cattle-cake and they improved livestock and the management of stock. Already, in the 1850s, the precursor of today's highly intensive animal production systems was in place in some parts of Norfolk: 'young stock would be kept under cover rather than in open yards both to keep them warmer and to conserve manure' (Martins 1988: 40). However, these advances in both livestock and cereal farming depended on the farmer being able to afford the high cost of purchased inputs of fertilizer and animal feeds. Intensive farming was only profitable in times of high prices. Many farmers on smaller holdings – in Norfolk as in Hooten Pagnell – continued to use crop rotations, relying mainly on animal manure to keep their soil fertile, and to feed root crops to their stock.

The English landowning aristocracy was powerful not only because of the range of its interests, but also because of the class structure it dominated. Unlike the situation in much of Eastern and Southern Europe, or the plantations of the American South, there was little class polarization in the English countryside between owners and labourers, but there was a graded hierarchy in which tenant farmers played a unique intermediate role.

The peculiarity of the English 'tripartite' system of capitalist agriculture, to which Marx gave so much attention in *Capital*, was that it freed its principal beneficiaries, the large landlords, to indulge their interest in anything that lay outside farming, including politics and business in many cases. Until about 1880, over half the ranks of the very wealthy in Britain were made up of landowners who were industrial entrepreneurs, and the hold of large landowners on British government continued certainly until the 1920s, and arguably later.

The degree of concentration of landholding in England was impressive in the late nineteenth century. In the 1870s no less than 80 per cent of private real estate was controlled by 7,000 people. A quarter of the land area was held in estates of more than 10,000 acres, owned by only 360 landholding families. In addition, smaller estates of up to 1,000 acres accounted for an additional 50 per cent of the land area. A large landowner like the Duke of Devonshire owned more land than the Kaiser. The Russian aristocracy, to suggest another comparison, owned just 14 per cent of the territory of European Russia. The English landed classes were prosperous, managed relatively productive estates, and many of them had long since made the transition to being industrial entrepreneurs, a transition which was to elude most large landholding elites (Barrington Moore 1969).

Before the First World War the *average* farm size in the United Kingdom was, nevertheless, much smaller than today: 63 acres compared with 278 acres in 1980. However, in the closing decade of the agricultural depression which continued until the First World War, many farmers had left the land and the number of small farms already had fallen dramatically. The agricultural labour force was only 10 per cent of total employment in 1891, but it continued to fall in the succeeding decades. The 'New Domesday Survey' of 1873 showed that 7,000 people owned four-fifths of the land but, according to Perrot (1968), four-fifths of these large estates had decreased in size a century later.

Most of this land changed hands in the inter-war period. Between 1918 and 1926 approximately 7 million acres of land changed hands in Britain, and in the same period owner-occupation of land rose from 11 per cent to 36 per cent (see

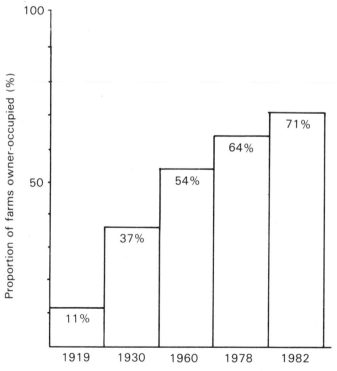

Figure 2.1 Owner-occupation of farms in the UK

Figure 2.1). Of the estates over 2,500 acres that survived these changes, most were of nineteenth-century origin, but most, as we have seen, were also considerably reduced in size. Of the 'smaller' landowners, the gentry and squirearchy owning between 1,000 and 3,000 acres in 1873, only about a third moved off their land. One estimate is that a century later some 1,500 families descended from the nineteenth-century gentry owned about 4 per cent of the land in England and Wales (Massey and Catalano 1978). The watershed in English agricultural history was the election of a Liberal government in 1906, which encouraged sales of freehold land and introduced measures to strengthen the position of tenants (Thompson 1963).

Unlike the landed nobility most of the landed gentry remained on their land, to farm it themselves, after ceasing to have tenants. This has given a special complexion to British landholding, and more continuity than in most other European countries. Owner-occupiers today are made up of three groups: former tenants who acquired their freehold, mainly in the 1920s or 1930s; newcomers to farming as the result of official government encouragement; and what remains of the former gentry, now comfortably-off capitalist farmers.

The role of the state in encouraging freehold farm ownership in Britain is an important one. It began with Lloyd George's Liberal administration, partly as a response to the depressed condition of agriculture, which had taken a heavy toll of tenants. Farmland was cheap between the wars. The average price of farmland in 1939 was £25 per acre (compared with £2,000 per acre in 1986). Heavy death duties had forced large estate owners to sell off part of their land and served to let new owners in. The main concern of inter-war governments was to increase output and productivity and to bring neglected land back into production. Mechanization was encouraged, although capital for it was scarce. The Agricultural Marketing Acts of 1931 and 1933 gave farmers a degree of concerted power over their produce, and by 1928 agricultural land and holdings were already relieved of rates. As we saw in the discussion of Hooten Pagnell, the first major reform had been to provide tenant farmers with more choice and security of tenure. The Agricultural Holding Act (1923) allowed sitting tenants greater freedom of cropping and much greater security from being evicted. The balance of changes already favoured the tenant over the landlord; the effect of government intervention was to make it easier for tenants to play an independent role and for some of them to acquire freeholds.

Wartime incentives to farmers shifted advantages still further in the direction of owner-occupiers, and established a system of price reviews and guarantees which assisted them, and reduced the risks which would have accompanied the transition from tenancy to owner-occupation had it occurred only a decade or two earlier. The wider implications of these changes are dicussed in Chapter 6.

Perhaps the most important aspect of changes in landownership in Britain is that increased owner-occupancy, and the break up of the 'Great Estates' has not reduced the concentration of land *holding*, although of course, it has distributed land *ownership* more widely. Two-thirds of all farms in England and Wales are less than 50 hectares in size, and are reliant, almost exclusively, on family labour (see Table 2.1). At the same time there has been a continued drift from the land since 1950. The total number of farm holdings has been halved and an average of 8,500 farms a year have ceased to exist. Almost a third of farms in the United Kingdom are between 50 and 500 hectares and these farms account for over two-thirds of the farmed land area. They appear to constitute the archetypal medium-sized family farm, most of which are mainly or wholly owned. Expressed differently, however, these figures imply considerable inequality: 16 per cent of the largest farms account for 58 per cent of the farmed land – a highly skewed distribution of land.

Table 2.1 Concentration of landholding in the UK 1982

Area of holding (hectares)	Proportion of total holdings (%)	Proportion of total area (%)
1–10	26.5	2.0
10–30	24.6	7.9
30–50	15.0	10.0
50–100	18.1	21.7
100–200	10.4	24.2
200–300	2.9	11.7
300–500	1.7	11.0
Over 500	0.8	11.5

Source: Agristats (1984: 156)

THE EUROPEAN DIMENSION

Just twenty years ago Henri Mendras expressed the view that the survival of the French peasantry was a *sine qua non* for the survival of French society itself:

> Most historians of agriculture (in France) have admired the *sagesse française* that kept the nation from pushing the agricultural revolution of the eighteenth century to extreme social consequences and enabled us to conserve a large peasant class, while the British, yielding to the logic of the industrial economy, sacrificed their agriculture to the development of industry.

> (Mendras 1970: 5)

Plate 2.1 French charcuterie displaying non-industrially processed food

Such sentiments still find expression in French thinking including that of political leaders. President Mitterand, speaking at a Chatham House conference in January 1987, is reported to have criticized 'purely industrial' agriculture as offering large agri-food factories but 'no peasants and farmers', a concept out of keeping with the Treaty of Rome's signatories, who had wanted 'a certain kind of rural civilization' (*The Times* 7 February 1987). The assumption, very clearly, is that by maintaining small family farmers on the land, Europe can create a bulwark against urban society and 'industrial' agriculture.

The history of West European farming since the Second World War has been one of social transformation, in which the pre-war peasantry has disappeared, to be replaced by a politically influential class of family producers, with clear sectional interests and a high degree of market integration. The transformation has been marked by widespread migration to the towns, and to urban employment; a decline in self-provisioning on the farm; the virtual disappearance of traditional systems of tenancy, like *mezzadria*; and the rise of co-operatives and syndicates, aimed at securing financial advantages from sympathetic governments.

In the European Community (EC) as a whole, more than half the agricultural area is farmed by the owner. In early 1990, in some countries (France, the Netherlands, the United Kingdom

Plate 2.2 Spanish peasant family farm, Galicia, Spain

and West Germany) between a third and a half of the agricultural area was tenanted, but in others (Denmark, Ireland and Italy) the overwhelming majority of the agricultural area was owner-occupied. Where tenants farm land in the EC today they have considerable security of tenure and controlled rents, making farm mobility difficult.

The contrasts between European farming today and that of twenty years ago are instructive. The first evident change is in the numbers employed in agriculture. Between 1960 and 1980 the farm population in the (Nine) EC countries was almost halved: from 17 million to 9.4 million. A similar reduction took place in the United States, where the farm population fell from 15.6 million (1960) to 7.5 million (1979). This dramatic decline, in the space of two decades, was associated with two processes: agricultural labourers leaving the land and a declining number of small family farmers. As in the United Kingdom, wage-labourers left West European farming more rapidly than farmers and their families. For example, between 1970 and 1975 the decline in the number of farms was 2 per cent per annum in the European Community of the Six, but the rate at which farmers and labourers left agriculture was 3 per cent per annum. Much of the loss was in smallholdings. In West Germany, for example, the number of farms fell from 1,600,000 in 1949 to 979,000 in 1989. Of those who remained in farming in 1980, over a quarter farmed holdings of more than 20 hectares, although a half of West German farms were still under 10 hectares. The changes were also more marked in some countries than in others; in Belgium, Luxembourg and West Germany farm numbers declined by over 50 per cent, in Greece and Ireland by barely a quarter. Furthermore, the accession of Spain and Portugal to the EC has served to increase slightly the number of very small farms.

The restructuring of the agricultural sector in post-war Europe was undertaken in a 'spontaneous' fashion, largely unconnected with policy measures to 'rationalize' farm structures. The much publicized Mansholt Plan (1968) of the European Community argued that European agriculture needed to be reformed from within, producing fewer, small holdings and more 'economically sized' farms. The Plan even specified 'ideal' sizes: 80–120 hectares for cereals; 40–60 cows for dairy farms; 150–200 head of cattle for beef production.

These ideas were incorporated within an important EC directive issued in April 1972, which urged a more selective use of grants and subsidies to enlarge the size of remaining farms. However, structural policy to reform farming has received much less emphasis than policies to ensure food security within Europe, and the maintenance of farmers' incomes. The EC directive to encourage elderly farmers to leave agriculture (Directive 72/160) has only been effective where labour market demand reinforced its provisions. In the 1960s under the French policy *Indemnité Viagère de Départ* (IVD), elderly farmers were offered a pension if they gave up their land. Between 1963 and 1973 some 400,000 French farmers took the 'retirement option', liberating about 7 million hectares of land and enabling half a million younger farmers to increase their farm size or join agriculture (Hamilton 1985: 348). However, there is little evidence that programmes of this kind speeded up what was already a 'natural' process, as older farmers, many of them without children interested in continuing to farm, left agriculture altogether. The majority of agricultural holdings are still farmed primarily with family labour. In most countries two-thirds of these farms are under 20 hectares in size. The major accomodation to change for small farmers in Europe has been in another direction: that of part-time farming.

If we define a 'part-time' farm as one 'which cannot provide enough work for one family' then 43 per cent of agricultural holdings within the Community are part-time farms (Frank 1983). This does not mean that all part-time holdings are necessarily farmed on this basis; many are farmed 'full-time' by poor and elderly people. The heavy reliance on family labour is also clear from 1975 data: of the 12.7 million people working in agriculture in the European Community, farm occupiers and their families accounted for 11.8 million. If we exclude casual hired labour, only just over a million people were regularly employed on farms for wages (Frank 1983). In 1975, only just over a third of all farmers were full-time although another 20 per cent of farmers spent more time on agricultural than on other work. This left 44 per cent of farmers for whom agricultural work represented a smaller proportion of their working lives than other forms of employment off the farm. Many farmers who work part-time were unable to secure other employment. Their farming activities thus represent a form of 'hidden underemployment', which in some countries like Spain (Etzexarreta and Viladomiu 1989) may be considerable. Data on part-time farming do not take account of social benefits such as pensions or voluntary work, which makes it difficult to calculate the real extent of underemployment among family farmers in the European Community. Certainly there is considerable geographical variation: in Sicily, and in most areas of *minifundia* in Spain and Portugal, most family farmers are underemployed for most of the year.

Information on part-time farming within the Community has another, more basic flaw: it does not record the numbers of spouses (usually wives) who gain income from non-farm employment. The 1960s and 1970s saw major shifts occurring in women's employment in rural Europe. There is some evidence that women in Europe worked longer hours in farming households (Franklin 1970) but their private entrepreneurial activities have been almost completely ignored. In 1960 over half of women 'dependents' in West German farming families were over 45 years of age (compared with only 21 per cent of male 'dependents'). Many of these women were 'gainfully employed' in non-farm activities, but the data are sparse. We do know, however, that in the mid-1960s 38 per cent of the food consumed by French farming families was still produced on the farm. It had been over 50 per cent a decade earlier, and was still over 60 per cent in West Germany in 1962. As the proportion of food consumed by the family from its own produciton fell, food had to be bought and income raised to buy it. Women's incomes helped to finance these food purchases, and those of other consumer goods.

Changing consumption patterns, linked to 'urban' lifestyles, have made major inroads in rural areas of Western Europe. The ability to buy more consumer goods, together with the expectation of greater affluence, obviously rests on increased cash income, and where opportunities to gain this from farm work are limited it is often women who have augmented the family's budget. Choices are also implied for the farm holding as an enterprise: perhaps between a new baler and a new car or a video recorder. The struggle of family farmers to survive is bound up with the flexibility of women's labour, which has enabled the farm enterprise to accomodate to the changing work demands of increased commoditization.

In this respect the role of women's labour in farming runs parallel to that of women in the urban economy, whose employment in the industrial sector served to alter the technological basis of housework, and, by degrees, the nature of the popular diet. Before broadening our analysis to consider some of the ways in which changes in agriculture have affected the countryside in general, we need first to examine the role played by capital in transforming the farm enterprise.

CAPITAL AND THE FARM ENTERPRISE

At the national – and international – level, the most vivid illustration of the power of capital to transform the conditions of production, and with it the natural environment, is the way that the 'technological treadmill' has forced farmers to adapt to new conditions, or leave agriculture altogether. The massive increase in agricultural surpluses, which has marked recent decades, has been associated with a decline in the numbers of farmers in the industrialized countries. As Cox *et al.* argue,

Plate 2.3 Modern irrigation in Castille, Spain

in their contribution to the discussion, 'public money has been poured into [the agricultural industry] resulting in massively increased output, yet the numbers of farmers has steadily declined and farm income problems remain' (Cox *et al.* 1989: 114).

National statistics to support this picture are not difficult to find. Between 1950 and 1985 wheat production in the United Kingdom increased fivefold, as more land was brought into intensive production that would otherwise have remained in other crops, or left as pasture. Increased production on larger and more specialized farms was associated with an increased dependence on other, largely input sectors within the food system. While the large farmers have been the principal recipients of most of the funds ploughed into agriculture, the input sectors, such as machinery, fertilizers and agrichemicals, have been the joint beneficiaries. Within two decades (between 1955 and 1975), the proportion of total land area farmed in England and Wales, which formed part of large farms (those of over 300 acres) rose from 27 per cent to 43 per cent. Large farms dominated the landscape in a way they had never done before, even in the pre-enclosure countryside. At the same time, the period since 1950 has been marked by a continuous decline in the numbers of farmers: the number of farm holdings in the United Kingdom was halved and the numbers of full-time farmers declined by a third. As we have seen, farm workers left the land in even larger numbers: the number of full-time farm workers in 1985 was only 121,000, compared with 717,000 in 1950, that is, only 17 per cent of the figure thirty-five years earlier. Today, when farms tend to be larger, the farmer and the family members constitute almost two-thirds of the farm labour force.

The discussion of the subsumption of production relations of the farm by capital has recently been given considerable empirical force by research at University College, London (Munton *et al.* 1987; Marsden *et al.* 1987 and 1990). Undertaking empirical research in three very different areas of southern England made it possible for the research team to develop a typology which more accurately reflected the considerable variation in types of enterprise. The research team were interested in the process through which individual families' control over the entrepreneurial functions of farming was being diffused and weakened, as the enterprise became more subsumed to capital. Following Goodman and Redclift (1985), the research team took the view that the competitive movement of industrial and finance capitals followed an uneven course; the effects were spatially, as well as socially differentiated. As some sectors of the food system were valorized by capital, and the labour process restructured, others were left virtually unchanged. A similar perspective, but one which takes an even broader view of social differentiation in rural areas than that prompted by agriculture alone, is discussed in Bradley (1985). In both cases the researchers constructed empirical 'measures' of dependence on external capital, which enabled a comparison to be made of farms using a consistent set of indicators. A typological matrix provided the central axis around which the model was developed, and made it possible to locate the indicators, conceptually. The degree to which ownership and/or control of the means of production on the farm were diverted away from the head of a single family household was taken as the principal measure of the degree of direct subsumption. The extent to which the farms depend upon the external relations of production, such as credit, technology and marketing, indicated the degree of indirect subsumption.

Looking at farm business at the level of the enterprise rather than at their contribution to overall production showed, as we might expect, a relatively low level of direct subsumption to capital. Larger farms, as we have seen, contribute disproportionately to total production, but their relations of production are more like those of other large businesses than family farms. The typology developed by the researchers distinguished between 'family labour farms' at one end of the continuum, where labour was exclusively drawn from the family and the external dependence on capital was relatively low, and 'family-owned farm businesses' at the other end, where the retention of a 'family' interest did little to distinguish the farm from any other large business.

The strategies of farm families in negotiating their subsumption to capital were highly divergent in each case, and partly reflected different ways of coping with the 'technological treadmill' effect. The possibility, for example, of expanding or diversifying the farm enterprise, was influenced by factors like the inherited pattern of land ownership and the quality of the land. Throughout all the study areas it was established that capital had penetrated farm production relations more effectively through indirect subsumption mechanisms, such as credit, technology and marketing relations, than through more direct methods. Direct ownership and management by capital of farmland, and the conversion of the farm labour process into one which only used salaried workers, was unusual. The research also established that off-farm sources of income and capital heavily influenced the choice of strategy. These extra-farm components of livelihood determined the ability of the family unit to reproduce itself, and the ability of the farm business to expand, albeit with family labour as its core resource.

Considerable analytical refinement has been brought to the study of the farm family by the work of Sarah Whatmore (1988). Whatmore argues that the discussion of family farming frequently overlooks the 'family' in favour of the production unit, and conflates the discussion of kinship with that of the farm household. The farm household is rarely equivalent to kinship relations, since people are invariably related to others with whom they do not share a roof! In essence, the household expresses that part of the farm family's existence which can be geographically located. As in the urban households discussed in Chapter 1, the location of the household is determined by co-residence and common patterns of consumption, relations which are mediated by power and social divisions within the family itself. Whatmore argues, persuasively, that the analysis of family farming has usually emphasized the 'occupational characteristics' of farmers (Galeski 1972; Shanin 1972) without distinguishing between the different members of the household on the basis of gender relations. The emphasis in the literature is on the peculiarities of family farming *whoever undertakes it*, largely in terms of occupational skills, such as farm management. The development of agriculture is then presented, within this literature, as a process which weakens the family's economic functions, as farming households become more integrated within wider institutionalized patterns of behaviour.

Whatmore argues that the problem with this approach is that it takes far too narrow a view of the 'economic functions' of the farming family, confining them to the occupational sphere. The truly remarkable feature of family farming, for which her research provides empirical evidence, is that as agricultural specialization increases, as it usually does on most farms, the burden of labour is shifted from the paid labour force to women. Specifically, economic specialization on the farm usually depends on women being able to undertake some, or all, of the tasks performed by men. The paradox is that as the farm, as an economic unit, loses its self-sufficiency, the members of the farm family become more dependent on each other, and especially on the activities of women. The parallels with urban households discussed in the first chapter are very clear; as in urban areas increased commodity relations demand greater complementarity of roles between household members, and this *increases* as women take more paid employment in the outside economy.

It is therefore tempting to raise the issue in a form similar to that in the urban debate; women within farm families appear to lose their control over the means of production as they shoulder the burdens of combining 'housework' (which on a farm can mean taking telephone calls for the 'farmer' and even doing the monthly accounts) with paid employment outside the family. There is a central paradox, then, in the survival of the family farm, which can only be explained by the 'bridging' effect of women's labour. It is that as the farm becomes more capitalized – moving along the matrix from a family labour farm to a family-owned business – women's labour has to become more adaptable, less specialized, more 'polyvalent'. As in the urban sector there is a shift in the labour performed by women, between two spheres of activity, within the home and on the farm, and in both of these spheres women's labour is more or less 'invisible'. Labour is performed in the household for the family, and on the farm for the household. In neither case does this labour undermine the viability of the family farm; indeed it becomes the means through which the family farm is reproduced. The development of the family-based labour process on the farm has facilitated the capitalization of the farm, without transforming the entire labour force to that of wage earners. Where women's labour has been retained, within the farm family and in seasonal agricultural work outside the 'formal sector' of employment, it has proved as necessary as it is invisible.

THE COUNTRYSIDE:
IDEOLOGY AND PRACTICE

In a series of influential papers and books Howard Newby sought to correct the erroneous but widespread view, current in the 1970s, that the British countryside was governed by a broad consensus of values which emphasized continuity and 'community'. Newby argued that rural areas of the United Kingdom had been subjected to two different sets of policies, and their effect was to leave the farm worker in a vulnerable position (Newby *et al.* 1978; 1979). On the one hand government policies to ensure cheap food had passed the burden of maintaining agricultural support to the taxpayer and, especially after entry to the EC, to the consumer. This cost was invisible, but the effects were clear: price support policies had helped the large farmer most, and only enabled small farmers to maintain production through increasing subjection to the 'cost–price spiral', which constrained their profit margins. In addition inflation in land prices had proved a bonus for landowners in the 1970s, a period of high inflation generally.

On the other hand, argued Newby, the effect of planning policy, since the Second World War, had been to free agriculture from development control. Farming had been given special status within the economy, and this status had not been seriously questioned by either major political party. A political consensus surrounded the view that preserving the countryside and helping agriculture were essentially the same thing: what was good for British agriculture was good for Britain as a whole. In 1968 legislation was passed which curtailed this extraordinary degree of freedom, to a limited extent, and introduced the idea of Structure Plans, which took account not only of land use, in reaching planning decisions, but also population changes, employment, transport and housing. Newby argued that the full potential of Structure Plans could not be realized, and that this was not the fault of planners themselves. The real problem was that farmers and landowners still held political power in many rural areas together with middle-class 'incomers', and they were unwilling to change the balance of power in favour of farm workers and other members of the rural working class.

Planning during the 1960s and 1970s had directed new

industrial development away from rural areas, in the main, narrowing the employment opportunities open to rural people. By reducing the competition for labour from alternative employment, rural wages were maintained at a low level for agricultural workers. The same observation applied to housing: farmers were unwilling to encourage building of council houses, which would compete with tied cottages, and during the post-war period the proportion of farm workers in tied housing had increased. Councillors in rural areas, many of them farmers, were also unwilling to lower council rents without increasing rates, an unpopular measure unlikely to command the support of ratepayers in the countryside, many of them landowners. One of the consequences of these tensions was that housing need in rural areas came to be associated primarily with visual criteria, rather than social need. Planning controls on housing were principally an instrument of social exclusivity, helping to define the character of villages, and preserve them from what were seen as the 'negative' effects of development. These policies were given further political support from what Newby described as the 'adventitious population' of people new to the countryside: second-home owners, rural commuters and retired middle-class people. These people sought what Pahl called a 'village in the mind', and having found it were determined to preserve the 'traditional' character of rural communities (Pahl 1970).

The 1980s was a decade of considerable social change for the United Kingdom, and rural areas were exposed to most of the forces of market restructuring and cuts in public expenditure, which came to be associated with Mrs Thatcher's first ten years as Prime Minister. The idea of 'community' still had ideological resonance, indeed for some people it proved an even greater inducement to live in rural areas than formerly. The media paid considerable attention to the notion of community, as evidenced by the extraordinary popularity of 'soap operas' such as 'EastEnders' and 'Neighbours'. Government legislation has also recently placed the onus for care on the community, rather than local authorities, although in rural areas, as we shall see, traditional kinship rarely functions as a safety net, in the way it did before the last war. The deregulation of rural transport opened up entrepreneurial possibilities for passenger transport operators, motivated by profit, who had previously been debarred (in the main) from competition with subsidized public transport services. Rationalization of schools and other public facilities was taken a step further; rather as rail services had been in the 1960s, it was now necessary to prove that a local need existed for services before they could be retained.

Many of the features of social change in the 1980s, in Britain as a whole, have been equally apparent in rural areas. Home ownership has increased, although rather less rapidly than in urban areas where there were more council houses to sell. One consequence has been a blossoming of 'Neighbourhood Watch' schemes, which enlist the support of neighbours in protecting private property. It has been remarked that such schemes as 'Neighbourhood Watch', together with other trends such as do-it-yourself, demonstrate that the force of exposure to the market has driven people into the family, rather than into the community. In effect, the community has been privatized, and people act collectively in defence of their property interests, rather than as an expression of shared social objectives. In the era of 'self-help', 'citizenship' and voluntarism, what the local state does not provide people need to provide for themselves, and the local state provides relatively little, especially in rural areas. For the middle class the rhetoric of 'community' is increasingly a selective rhetoric. It refers to community associations (something that would have been a contradiction in strict *gemeinshaft/gesellschaft* terms) to the return to private interests of acting collectively. In rural areas, as in urban and suburban ones, people join sports clubs and special interest groups so that they can associate with like-minded people and pursue shared interests. Rather than attempt to widen their own social horizons, it can be argued, modern group membership – in rural areas as much as in urban areas – is an expression of status and class membership, rather than community affiliation. People in rural areas construct, through their recreational activities, formal networks of interaction, which serve to emphasize what they have in common. As we shall see, social divisions in rural areas are at least as great as in urban ones, so it follows that group membership reflects social differentiation in the society as a whole.

The way that rural tradition, notably the idea of 'community', has been worked and reworked, especially by the media, represents an important facet of ideological imagery in the United Kingdom today. As a recent discussion makes clear, it is important to establish which social classes *benefit* from this tradition, and which classes are responsible for *making* and *remaking* tradition in modern society (Hamnett *et al.* 1989). During the late 1970s and throughout the 1980s a new class of emerging managers and entrepreneurs, the beneficiaries of Tory economic policies, were able to associate themselves with some of the 'enduring' features of rural life, through acquiring the goods (most of them environmental) which typify the British countryside, such as land, period housing and shooting rights. In establishing themselves as the 'natural' inheritors of the traditional landed classes they gained legitimacy, and by occupying the countryside they converted a private fantasy into a reality. They were also able to develop their own consumer tastes for country life.

It would be wrong, however, to dismiss the re-creation of 'countryside ideology' simply as evidence that the countryside had become a 'theme park', in which most of the people were, in Pahl's phrase, simply 'props on the rural stage'. Recent decades have also seen an enormous growth of serious interest in conservation, in wider environmental values, that cannot be reduced to ideological refractions. As we shall argue in a later chapter, while the countryside has been contested by different social groups, each possessed of their image of what it is, a new environmental politics has sprung up, which appears to be answering wider political and social objectives, making the Green agenda a rich new vein of political thinking.

Recent discussion of social deprivation in the British countryside has underlined the bias in considering rural areas as bastions of tradition and continuity (Lowe, Bradley and Wright 1985; Bradley 1985; McLaughlin 1986). At the same time this discussion also helps throw light on the vexed issue of what actually constitutes a 'rural' society. As we saw at the beginning of this chapter, in the discussion of rural depopulation, designating an area as 'rural' is increasingly problematical, since so much of lowland England consists of urban 'dormitory' settlements. The discussion of rurality is important in another sense, however, for underlying the 'rural' designation is a frequently concealed suggestion that spatial, rather than social and economic factors, lie at the heart of social deprivation in the countryside.

The discussion of social deprivation in rural areas has been subjected to competing interpretations. First, there is the view that social deprivation in rural areas arises from the spatial characteristics of these areas: their relative isolation, the dispersal of population and the difficulty in identifying 'central places'. In these terms social deprivation reflects the problems *of* rural areas, rather than the problems *within* rural areas. The challenge facing planners, then, is to improve geographical access to services, seeking to equalize access between different local government areas. This view has tended to reinforce a *planning paradigm* (Bradley 1985) of rural development, which looks at the effects of economic and planning decisions, most of which have been taken beyond the confines of the village (Shaw 1979; Moseley 1980).

Another perspective is provided by what Lowe, Bradley and Wright (1985) describe, rather confusingly, as the 'sociological paradigm'. Drawing on the work of sociologists and social policy professionals, notably Runciman (1966) and Townshend (1979) as well as Newby, this approach places the emphasis on deprivation as the consequence of social inequality. Structural inequalities, found in rural as well as urban areas, are the principal determinants, in this view, of the pattern of social deprivation. Whereas the 'planning paradigm' was developed as an adjunct of policy making – which helps explain the emphasis on central place theory, and economies of scale in service provision, for example – the 'sociological paradigm' begins by examining the structural, class-based development which produces the problems that planning is supposed to address. The planning paradigm tends to gloss over the basis of social inequality *within* rural areas, on the assumption that rural areas are distinguished by more mutual aid and reciprocity, an assumption that is not supported empirically. Indeed, the normative commitment to 'village life' or 'community' that underlies the planning paradigm is rarely examined closely. As Bradley puts it, 'village England rhetoric . . . is pure ideology. On the one hand it masks social divisions in the village behind the banner of 'community'; on the other hand it consigns poverty, through moral rectitude, to the city' (Bradley 1985: 24–5).

The evidence of social deprivation in rural areas suggests that the dominant view of the British countryside as exhibiting social cohesion and consensus values is highly misleading.

As we have seen, official policy relates rural deprivation to a decline in service provision in rural areas. In this view rural deprivation stems from what Shaw (1979) called 'opportunity deprivation'; access to services is poor, these services can only be provided at high cost, and this cost is prohibitive. It is not that some rural people are deprived, it is that the deprived are rural people. There are difficulties with this line of argument, apart from its flawed logic, in that a common response to the assertion that deprivation is rooted in 'rurality' is to sidestep the argument altogether, by suggesting that rural people cannot share a condition of deprivation ('an urban concept') with urban people since it is held that they are more resourceful, more independent and more able to elicit help from friends and kin. The 'planning response' to rural deprivation is not, then, simply inadequate, it is actually misfounded.

As Bradley (1985) and McLaughlin (1986) have shown, with reference to data from a sample of five very different rural areas in England and Wales, the idea that extended families and solidaristic community groups exist in village England, providing support for the rural poor, is at best misconceived and at worst an ideological smokescreen. The effect of demographic restructuring since the last war has been to produce a breakdown in the supportive effect of close kinship ties. In rural/urban dormitory areas kinship networks have been transformed by the immigration of newcomers, who occupy much of the best housing and employment and who have greater geographical mobility in gaining access to services. Over half the people sampled by Bradley had no family living close to them in this kind of rural area.

In 'deeper' rural areas, the upland areas of England and Wales, and especially the North of England, local kinship networks were slightly stronger, but had been seriously disrupted by the immigration of retired people, who had also had a profound effect on the availability of local housing stock. This mobility of retired and 'second-home' owners into more isolated areas had also served to cut off local people from their kin, although the process was less marked. In all five study areas one of the most exposed groups was made up of single elderly people (especially women) of whom between 45 and 85 per cent (depending on study area) were living below the 'supplementary benefit' level.

81

Within this context it is important to ask whether 'community care' is a real possibility in rural areas. In Bradley's words:

> Thus, at one extreme social spaces are produced where isolated, elderly newcomers reside within close-knit, small-scale labour (markets) dominated by local familism. At the other end of the continuum, an older and more dependent, indigenous population has become isolated within a belt of relatively affluent, mobile newcomers, who have little . . . commitment to the locality.
>
> (Bradley 1985: 168)

In aggregate these surveys show that, using Government criteria based on Supplementary Benefit calculations, about a quarter of rural households are living at or below the margin of poverty . This group, since it is made up of large numbers of single, elderly people, constituted rather less – about a fifth of the rural population. Social deprivation in rural areas is widespread. Second, income inequality is greater in rural areas than in urban areas. In the rural areas sampled, 27 per cent of households had gross disposable incomes at least three times that of Supplementary Benefit levels, compared with a national average of only 11 per cent with incomes this large. In addition, the highest incomes in rural areas were 33 per cent higher than the national equivalent, while the lowest incomes were lower, on average, than the national average. As Bradley points out, this means that relative deprivation, if not absolute deprivation, is greater in rural areas than in urban ones (Bradley 1985: 152).

On the basis of all other available evidence, social deprivation in rural areas of England leads us to question the ideological construction that is placed on the 'countryside' (McLaughlin 1986). The hourly rates of pay for manual workers are well below the national average. Fewer retired people in rural areas receive additional pensions, in addition to the state minimum. Invisible earnings, often regarded as part of the prerequisites of agricultural labour, are much more important for the well-off than for the poor. In the form of free life insurance, medical cover, company cars and occupational pension schemes, such invisible earnings actually contribute to inequality in rural areas, since most are received by the relatively wealthy. Similarly, there is a higher proportion of unfit housing in rural areas than in towns, despite visual appearances to the contrary, while the relatively high incidence of outright ownership of houses in rural areas means that access to homes for 'first-time buyers' is poorer. This is especially marked in areas where migrants make up a sizeable proportion of the rural population. Similarly, the preponderance of elderly people among the (small) council house population in rural areas, has made the management of publicly-owned housing very difficult. Not surprisingly, rent-free tenure looms large in rural areas of England, farmers' tied cottages proving the most common example. Two-thirds of the households in tied accomodation were manual working class, and most of these households forfeited their home when they left their job.

Finally, access to shopping and other services demonstrates that poverty within rural areas, rather than between rural and urban areas, is the main feature of social deprivation in the British countryside. Two-thirds of elderly householders who shopped locally had no car (although car ownership generally is higher in rural than in urban areas). These were the people most exposed to higher prices and most threatened with the closure of local shops in villages. Half the households who reported having difficulty getting to know their local doctor had no car. Among the least mobile in rural areas, women fared worst. Not only were they less mobile than men, they were also less mobile than women in urban areas. Taking this evidence together it is difficult to disagree with Bradley that 'looked at from a perspective of relative poverty, the gulfs that exist between the life-chances of neighbours (in rural areas) are breathtaking' (Bradley 1985: 170).

DIVERSIFIED LIVELIHOODS: SPECIALIZED LAND HUSBANDRY

This chapter has argued that family farming in the industrialized countries has survived through being transformed. The last half-century has been marked by a huge contraction in the numbers of people employed in agriculture in North America and Western Europe. In the United States the number of farms in existence in 1990 was little more than a third what it was in 1930. Twenty-four million workers left agriculture between 1930 and 1980. In the EC the reduction in the number of farms was slightly slower, and certainly more delayed. Nevertheless, between 1950 and 1990 the number of farms will have been halved in the EC countries. This contraction in numbers does not imply a diminution of political influence. Farmers' groups have exerted considerable political influence on national policy and, until recently, their legitimacy has not been called into question. In explaining this process, which has helped to enshrine the family farm in political terms while effectively divesting it of some of its economic activities, we need to refer to three interrelated processes.

First, family farmers have diversified their livelihood strategies in ways that effectively reduce the impact of the restructuring process and have enabled them to reach a new accommodation with the wider polity. Today, many small farmers, perhaps a majority, obtain as much income from non-farm activities as from the farm itself. This proportion increases if we count women's employment off the farm, and the effects of pensions and other state benefits. Family farming has survived through non-farm activities without losing the source of livelihood (farming) that determines the survival of the farm. In addition, women's labour, both in the household and on the farm, has provided a flexible and easily mobilized response to change, enabling some family farms to develop as business enterprises, while others survive through intensive use of family labour, through a process of 'involution'.

Second, family farming has survived through more specialized land husbandry, which has served to link the farm with industrial suppliers 'upstream' and 'downstream'. As production has been further commoditized, the farm has ceased to operate as a food system in its own right, and the farmer exerts less effective entrepreneurial control over the management of the farm. The 'mixed farm' found in Hooten Pagnell in the early 1930s has not completely disappeared in England and Wales, but there are fewer such holdings than there were fifty years ago. More specialized land husbandry has served to relocate activities off the farm, and forced greater economic integration on farms. However, the ability to make this kind of accommodation to the market is dependent on disguising or ignoring contradictions between the needs of the family, and particularly those of women.

Third, and linked to both the above processes, is the increase in freehold ownership which has served to underwrite the contractual relationship between the family farm and the state. In the agricultural sector today, and in rural society as a whole, there is less direct confrontation between capital and labour, as labour markets have contracted in size and capital has been located outside the landholding unit. In England, the break-up of the 'Great Estates' induced a process of 'democratization' through which declining numbers of farmers were able to use their market power, assisted by the government, to acquire their freehold. On the continent of Europe the moves to freehold ownership which Weber observed in the nineteenth century were consolidated in the twentieth century. In North America, the 'family farm holding', for so long a bulwark of Jeffersonian democracy, survived the Depression and came to exert a tangible control on the levers of power, at least until the 'bubble' burst in the 1980s and the full extent of agricultural protection was recognized. Farmers in all the industrialized countries have exerted more influence as a sectional group than they ever exerted within the geographical and social confinement of *gemeinschaft* rural society. Increasing proprietorship became linked to new political strategies, and farmers have increasingly operated through co-operatives or producer groups. The effect of their closer incorporation in the industrial food system was to gain the assistance of the state in extending guarantees which established their very conditions of existence. The effect of credit, tax relief, rate relief and contract buying was to encourage farmers to seek enhanced participation in the food system, rather than to resist greater incorporation.

This process of accommodation to the modern food system is, increasingly, matched by one of 'resistance' from some groups of farmers, environmentalists and consumers. This is explored in Chapter 6, where the issue of the sustainability of modern agriculture is discussed more fully. That sustainability should attract so much interest in the context of agricultural change should not surprise us; it is intimately linked to the role played by capital in the expansion of the food system. Unlike the situation half a century before, concentration of capital (much of it outside farming) assumed importance in the post-Second World War period, rather than simply concentration of landholding. While the upper strata of family producers increased their capitalization and forged an alliance with agro-industrial capital, poorer, more marginal farmers 'resisted' in the only way open to them: by leaving farming altogether or by finding their major source of income outside, as 'part-time' farmers.

Rural society also needs to be examined in terms of the broader implications of these trends in farm ownership and concentration. In Chapter 6 we shall return to this question, by reviewing the implications of farm support for sustainability, and the conflicts between environmentalists and farmers. As we have seen, social deprivation in the English countryside has been obscured by references to 'rurality' and 'community', images which are often imbued with a romantic flavour. The countryside has, meanwhile become an increasingly important area of consumption, and one that can assume even greater ideological significance within a rapidly changing urban society. The paradox with which we are left is that while modern urban society has elevated 'rural' values to the ideological level , the society from which they are supposed to emanate has passed away. In this sense the passing of rural society might be seen as a *sine qua non* for the development of the modern food system.

The next chapter examines the way in which the agri-food system developed in the United States, during and after the New Deal period. The agricultural/technology policy model which resulted, served to link farmers more closely to both 'upstream' and 'downstream' industries, and became central to the food system's development. This model, as we shall see, rests squarely on a wider market for food products which incorporate additional 'value added' in their production. The technology to achieve this policy goal developed together with a greater integration of consumers and producers, and a more differentiated range of food commodities. First, however, we need to examine the 'motor' of the modern food system, assembled around farm support and the diffusion of new technologies which came into being in the United States.

3

THE ORIGINS OF THE MODERN AGRI-FOOD SYSTEM

This chapter examines two themes which interact, like major and minor chords, to determine the structure and development of the modern agri-food system. The 'major theme', played out at the intersectoral level, is the role of agriculture in the expansion of industrial capitalism, that is, within the general process of capitalist accumulation. The central argument is that, both in the extensive and intensive or 'fordist' regimes of accumulation (Aglietta 1979), the role of agriculture has been to relieve downward pressure on the rate of profit by furnishing staple foodstuffs or 'wage goods' at low real prices to the urban industrial sector.

The extensive regime, which prevailed roughly from the mid-nineteenth century until the First World War, is characterized by the concentration of accumulation and technological change in the heavy producer goods industries, relatively slow productivity growth, competitive markets, and the predominance of non-capitalist commodity relations in the mode of consumption and the domestic sphere (de Vroey 1984). As we noted in the discussion of the gender division of labour and food preparation in the urban household in Chapter 1 and the farm labour process in Chapter 2, non-commoditized domestic labour was central to the reproduction of the urban labour force and the family farm before the Second World War. Characteristic features of the intensive regime of accumulation, which was consolidated after 1945, include the interdependent expansion of capital and consumer goods industries, rapid technical change and productivity growth, monopoly market structures, the transnationalization of production and exchange, and mass consumption of standardized commodities. Commodity relations penetrate all spheres of consumption as use values are commoditized and mass-produced, leading to the concomitant decline of domestic labour and non-capitalist goods and services. As we saw in Chapter 1, the increasing commoditization of the domestic labour process associated with new food preparation technologies and the diffusion of 'white goods' has transformed women's working lives inside and outside the home, 'freeing' them to enter the labour market. These changes are an integral, catalytic element in the rise of the modern food system, opening up new consumer markets for processed or 'industrial' food products, including the expansion of catering and fast-food outlets, which have revolutionized what, how, and where we eat.

The 'classical' role of agriculture as a source of cheap wage-goods persisted in the transition to 'fordism' in the inter-war period but it then acquired a new significance. With industrial real wages institutionally linked to productivity gains via collective bargaining, the goal of cheap food policies now was to release a widening margin of purchasing power for non-food consumption, reinforcing the integration of the industrial working class in the market for mass-produced goods and services. The maintenance of inter-sectoral terms of trade favourable to the urban sectors remained important throughout the inter-war period, and constituted the principal aim of state agricultural policy in supporting the accumulation process. This continuity in macro-level objectives can be seen in the development of New Deal-type programmes in response to the accumulation crisis of the 1930s. This improvised response to agricultural surpluses subsequently become the model for new state–agriculture relations during the 'golden age' of intensive accumulation after the Second World War.

The 'minor' theme, though of fundamental importance for environmental change and therefore central to this book, also focuses on accumulation, but within the agri-food system rather than within the economy as a whole. It is argued that the dual biological constraints represented by the agricultural production process, together with the physiological requirements of human consumption, have determined *sui generis* patterns of accumulation in agri-food sectors. The agri-food system does not constitute merely another branch of manufacturing, and its transformation under industrial capitalism is not reducible to the case of handicraft activities or domestic industry. The industrialization of the agri-food system has been analysed using the concepts of appropriationism and substitutionism, which explain these specificities and the historical development of agro-industrial capitals (Goodman *et al.* 1987). As we shall see, the incomplete transformation of agriculture by these capitals, and the consequent fragmentation of innovation, has resulted in a production process whose consequences are highly problematic for the natural environment, nutrition, and the sustainability of agricultural production itself. Industrial capitals pursue independent sales and R&D strategies to promote wider use of their inputs, disregarding their impact on what Marx called the conditions of production or external nature. This disregard for the separate and cumulative effects of agro-industrial innovation on both the immediate production base and the rural environment at large is manifest in habitat destruction, monocultures, varietal specialization, soil erosion, the pollution of water supplies, etc. The common denominator of this degradation is unfettered capitalist accumulation, whose effects are felt not only within agriculture but also in the production and consumption of commodities throughout the food system.

The dynamics of accumulation at the sectoral and aggregate levels are the recurrent themes of this chapter as industrial capitalism moves from extensive to intensive development. In agriculture, this transition is marked by the sustained growth of productivity whose cornerstone is provided by revolutionary advances in plant genetics. At the wider level, the directions taken by the food system within the regime of intensive accumulation are shaped by the interaction between cheap food policies, which create distinctive patterns of development, and the constant drive of industrial capitals to exert greater control over 'nature' in agricultural production, food processing and manufacture. The post-war expansion of the food system is linked to two processes: the need to develop new markets for commodities and labour in industrial societies, and the potentiality of bringing science to bear on the technology of agriculture and food production. In this chapter, we present an analytical framework for understanding innovation in the agri-food system, and then go on to consider the integration of agriculture in extensive and intensive regimes of industrial accumulation.

ACCUMULATION AND INNOVATION IN THE AGRI-FOOD SYSTEM: AN INTERPRETATIVE FRAMEWORK

To understand accumulation in the agri-food system, it is essential to recognize that industrialization has taken a markedly different course from other production systems. The biological production–consumption cycle underpinning this system has imposed distinctive constraints on the technical and social organization of production. Whereas the general tendency in non-food industries has been towards growing independence from renewable primary commodity inputs, with the chemical industry replacing agriculture as the principal source of raw material feedstocks, the modern food industry has remained on the margin of this transition from the biosphere to the geosphere. With few exceptions, such as soya beans, the supply structures of the food industry continue to be dominated by the same agricultural commodities which were prominent in the pre-industrial food system. In similar fashion, as we saw in Chapter 1, many activities incorporated in the labour process in food manufacturing were transferred, almost directly, from the home, and performed almost universally by women. As a result of this biologically determined rigidity in demand and supply structures, the food industry has developed around specific agricultural products to form specialized food chains linking farm and table. Even after decades of diversification from their original product base, industrial capitals in food processing, such as Nestlé and Unilever, can still be identified by their specialized historical location in the food chain. The prominent role of product differentiation in food marketing strategies is a further reflection of this rigidity. Only in the 1980s with the advent of biotechnologies are these linear structures beginning to weaken, as we shall see in Chapter 5.

It has been suggested elsewhere that the distinctive patterns of innovation in the agri-food system, which set it apart from other industries, can be analysed in terms of 'appropriationism' (Goodman et al. 1987). The biological production–consumption cycle has resisted direct and *unified* transformation by industrial capitals, leading to partial and historically discontinuous appropriations of rural production activities. These separate appropriations 'define the origins of agro-industrial capitals and the complex of equipment, processing, seeds, and agri-chemical sectors' (Goodman *et al.* 1987: 7). Biological constraints therefore have led to innovations by capitals which, strategically for the historical development of agriculture, have been located *outside* the immediate production process. As individual rural activities are appropriated, with tractors replacing animal power and synthetic chemicals taking the place of manure, for example, the individual sectors so established become *autonomous* sources of innovation. The agro-industrial 'complex' thus has grown by historical accretion, adding new sectors as industrial appropriation progresses. The key point to recognize is that biological constraints (photosynthesis, gestation, species diversity, land as space, etc.) have led to the *fragmentation* of the innovation process, whose origins and dynamic lie outside direct agricultural production (Goodman and Wilkinson 1990). This fragmentation helps to explain the atomistic structure of agriculture, centred on family labour forms of production, certain characteristics of state intervention, and the degradation of the rural environment. As we shall observe in Chapter 5, the reinsertion of these fragmentary, industrially produced inputs into nature poses problems for environmental adaptation.

The physiological requirements of human consumption similarly have constrained the efforts of industrial capitals to substitute industrial products for agricultural food and raw materials. Nevertheless, the expansion of the food industry arises from a progressive tendency to achieve the industrial production of food, whether by reducing the rural product to a simple industrial input or eliminating its use entirely. This process of 'substitutionism' embraces the increasing utilization of non-agricultural raw materials and the creation of industrial substitutes for food and fibres, with the chemical industry as its paradigm. Historically, this tendency began by breaking the direct links between field and table, producer and consumer. Once mechanized food processing had been introduced, food products could be differentiated not simply by natural qualities but also by their industrial attributes, brand names and other merchandising techniques.

> With this foundation, which was established in its essentials between 1870 and 1914, the food industry subsequently could turn its attention to effecting qualitative changes in the organic composition of food and in the general perception of what constitutes food. . . Industrial capital thus embarked on the path that was to transform consumption patterns, distribution systems and the domestic routines of daily life.
>
> (Goodman *et al*. 1987: 60)

In fresh or perishable foods production systems, such as fruit and vegetables, industrial reorganization has taken a different course. Briefly, this is characterized by the combination of adaptive genetic selection, geographically diversified supply sources, and large-scale industrial operations, including direct production in such cases as bananas, lettuce and processing tomatoes. The main aims have been to standardize product quality and minimize seasonal fluctuations in an effort to approximate conditions of continuous production.

We have suggested that biological constraints in agricultural production generally have impeded its direct and unified transformation by industrial capital. Irrespective of the form of production in modern agriculture, whether family farm or corporate enterprise, the technical base of the production system is intrinsically the same. Apart from notable exceptions in livestock production and horticulture, industrial capital has been unable to revolutionize the agricultural production process to place it beyond the technological and organizational capacity of the family producer. Although there has been a persistent increase in farm size, family forms of production have accompanied this trend without conceding *technological* superiority to large wage-labour enterprises. These relations of exploitation are capitalist but the labour process retains recognizable links with the inherited, artisan-based pre-industrial activity. Direct producers continue to bring their individual blend of skills, experience and judgment to bear on the organization of the immediate production process. At the same time, despite the traditional *appearance* of agriculture, technological change has developed rapidly, particularly with the apparatus of government support established since the 1930s. Indeed, the social transformation of agriculture is due less to the revolutionary nature of innovation *per se* than the institutional incentives to early adoption and technological competition introduced by the state. In the post-Second World War period, for example, the consolidation of farms and the rural exodus experienced in the United States and Western Europe are mainly the result of the cost–price squeeze exerted by the state-subsidized technological treadmill. In large part, social differentiation and restructuring have been politically engineered rather than the result of ineluctable technological advantages of scale in production.

We turn now to examine the articulation of agriculture with industrial capitalism under the extensive regime of accumulation, when the real price of food was the principal determinant of the reproduction cost of labour.

EXTENSIVE ACCUMULATION AND THE EXPANDING AGRICULTURAL FRONTIER

The extensive regime of accumulation, in which the industrial labour force constitutes a cost to capital rather than a market for manufactured consumer goods, broadly coincides with the period from the mid-nineteenth century to 1914. For our purposes, the distinctive features of the mode of consumption under this regime are that staple foodstuffs and non-commoditized domestic labour engaged in producing means of individual consumption, including food preparation, have fundamental importance in the reproduction of the labour force. Aglietta (1979) argues that capitalist industrialization provokes a dual structural change – 'a separation between labour power and means of production. . . and a destruction of the spatio-temporal environment moulded by pre-capitalist forms of production' (Aglietta 1979: 80). However, these two elements do not necessarily coincide. Indeed, for a certain period, capitalism may benefit from the partial establishment of the wage relation; that is, 'from the reconstitution of labour-power by a non-capitalist environment in which it is inserted' (1979: 80). There is then 'an incomplete separation between productive and domestic activities, and . . . *a domination of non-*

93

commodity relations over commodity relations in the mode of consumption' (1979: 80). The practical implications in terms of the physical drudgery of the working lives and daily routine of millions of working-class women – unrelieved by the mass-produced 'white goods' of the fordist mode of consumption – were vividly described in Chapter 1.

This traditional 'social environment' or 'way of life' is destroyed in the leading capitalist countries in the later nineteenth century by the development of heavy industry, which 'enforces the total uprooting that is characteristic of the wage relation: the separation of labour-power from all its conditions of existence. The mode of life of the wage-earning class then suffers a deep degradation' (Aglietta 1979: 81). Aglietta also suggests that

> The logic of this structural transformation is the production of a new mode of consumption expressing the complete realization of the wage relation. This mode of consumption is characterized by the domination of commodity relations over non-commodity relations. . . Capitalism can reproduce itself only by an incessant accumulation which develops as a mass production and consumption of commodities, a phenomenon generalized to embrace the sum total of activities of social life.
>
> (Aglietta 1979:81)

The destruction of 'traditional ways of life' and the commoditization of consumption thus prepare the way for the regime of intensive accumulation in which the reproduction of labour power itself is fully integrated with capitalist expansion. This is getting ahead of the analysis, however.

In the period before the First World War, industrial capitalism was consolidated by its expansion in overseas markets in the periphery and the colonial possessions of the leading powers. Merchant capitals as 'the agent of industrial capital' (Kay 1975) concomitantly extended the frontiers of agricultural production in the temperate grain-producing areas of the New World, profiting from the revolution in transport and communications: the railway, the telegraph and the steamship. The spatial restructuring of the world economy was the principal mechanism in readjusting agricultural production to the demands of industrial growth for staple foods and raw materials. This new international division of labour with Britain as its axis overcame the resource constraints facing industrial capitalism in Western Europe by intensifying commodity production and export specialization in regions of recent settlement, including the United States, the Dominions and the River Plate countries.

The gradual formation of a world grain market after 1850 effectively resolved the 'agrarian question' in Western Europe by ensuring cheap supplies of staple foodstuffs for the expanding urban labour force. Indeed, the collapse of world grain prices in the mid-1870s undermined large-scale capitalist arable agriculture in Europe, notably in free-trade Britain, and initiated the long pre-eminence of family labour forms of production in the advanced industrial countries. Similarly, a variety of different forms of production – small commodity producers in North America, sharecroppers in Argentina and peasants in Eastern Europe – contributed to this growth of world output, which was achieved largely by extending the margin of cultivation to incorporate new land. This extensive spatial integration of grain production systems underlines the inability of industrial capitals to revolutionize the agricultural production process directly.

Macro-level intersectoral relations between agriculture and industry thus were redefined by the nineteenth-century expansion of the world economy, which destroyed the locational dependence of Western European industry on *domestic* agriculture. During the 'hungry forties', industrial accumulation was still vulnerable to what Hobsbawm (1979) calls 'the ancient agrarian cycle'. The emergence of a new, more complex international division of labour after 1850 meant that local harvest crises had a diminishing impact on economic activity, as well as dispelling the spectre of famine and food riots which had haunted the European ruling classes in the 1840s.

> Moreover, as the world economy strengthened its grip, even the fortunes of agriculture were to depend much less on the fluctuations of nature than those of world prices – as the great agrarian depressions of the 1870s and 1880s were to demonstrate.
>
> (Hobsbawm 1979:70)

The real significance of the new world division of labour is that it extended 'the resource base of the western world' (North 1958: 557). The hold which the rural landed classes potentially could exert over industrial accumulation was broken by the settlement of new agricultural frontiers in the New World and the Dominions. Their economic and political power was dealt a decisive blow by the protracted 'agrarian crisis' unleashed by the flood of New World grain imports after 1870. With the agricultural supply of strategic wage goods – grains and meat – organized on a world scale, *national* agricultures in Western Europe no longer represented a potential constraint on industrial accumulation. They retained importance as reserves of labour power, as internal markets for manufactures and the emerging agro-industrial sectors, and as bulwarks for bourgeois regimes. As Hobsbawm observes, 'What a growing part of agriculture all over the world had in common was subjection to the industrial world economy' (Hobsbawm 1979: 190).

With the trade cycle replacing 'the ancient agrarian cycle' as the decisive determinant of economic fluctuations, the repercussions of the unstable 'feast or famine' pattern of output were increasingly internalized within agriculture and lost their *systemic* importance. With simultaneous access to geographically separate production zones, the formation of the world market freed industrial capitals from the seasonality of individual national agricultures, approximating the continuous production process characteristic of industry. Furthermore, world market forces exerted constant downward pressure on farm incomes since local scarcity and high prices were easily relieved by imports, while gluts in the major producing regions rapidly depressed world prices. For the first time, over-production assumed structural proportions as new agricultural frontiers were incorporated into the world market, creating the foundations of the 'farm problem' in Western Europe which defies solution to the present day. Hobsbawm (1979) estimates that the area of cropland rose by 250 million acres between 1840 and 1888, with roughly half of this increase in the United States, where the cultivated area trebled.

The full force of this expansion, temporarily delayed by the Crimean War and the American Civil War, was felt after 1870 during the Great Depression, when world grain prices collapsed. With the exception of Britain, the overwhelming challenge of overseas supplies provoked the 'first wave of protectionism' in Western European agriculture (Tracy 1982). European responses to the catastrophic decline of world wheat prices are summarized by Kindleberger:

> In Britain, agriculture was permitted to be liquidated. In Germany large-scale agriculture sought and obtained protection for itself. In France . . . agriculture . . . successfully defended its position with tariffs. In Italy the response was to emigrate. In Denmark grain production was converted to animal husbandry.
>
> (Kindleberger 1951: 37)

EARLY INDUSTRIALIZATION OF THE AGRICULTURAL PRODUCTION PROCESS

The extensive regime of accumulation in industry and trade was accompanied by the emergence of specialized agro-industrial capitals, notably those drawing on technological advances in mechanical engineering and industrial chemistry. The later nineteenth century also saw the beginnings of large-scale industrial food processing – sugar-refining, automatic roller-milling for wheat flour, powdered milk, canning and refrigeration – and food manufacture. Before 1914, the historical discontinuities of industrial appropriation had already established agricultural machinery and fertilizers as powerful, though separate, branches of agro-industry. These capitals followed relatively independent trajectories before the introduction of hybrid seeds, with mechanization advancing rapidly in the United States, where plentiful land and scarce labour made the human and animal energy base of the labour process the logical focus of industrial appropriation. In Western Europe, on the other hand, more abundant labour and the relatively fixed supply of land led to greater concentration on the biological constraints to higher crop yields, particularly control of 'the nitrogen economy'. The growth of the heavy chemicals industry in the nineteenth century was built mainly on the appropriation of the plant nutrient supply, initially through the demand for sulphuric acid which was used to produce superphosphates. This trajectory continued with the search for efficient industrial processes to fix nitrogen, culminating in the ammonia synthesis and the rise of the nitrogen fertilizer industry in the 1920s. The abundance of synthetic nitrogen, whose real price fell dramatically as industrial capacity expanded, together with hybrid seeds, provided the foundations for the transition to more intensive agricultural systems and higher crop yields after the early 1930s.

These genetic innovations were preceded almost a century earlier by the mechanization of the small grains harvest in the United States – a classic example of industrial appropriation. A series of mechanical innovations after 1840, including ploughs, seed drills and reapers, replaced manual power and led to the dramatic growth of labour productivity. This growth gained further impetus with the introduction of tractors and combine harvesters, although their wide diffusion really belongs to the inter-war period. According to Rasmussen (1962), man-hours per acre of wheat declined from 56 in 1800, to 35 in 1840, to 20 in 1880 and to 15 in 1900. Similarly, Parker and Klein (1966) estimate that output per man-hour in US grain production increased fourfold between 1840–60 and 1900–10. Significantly, these falling labour requirements put the scale advantages of mechanization within reach of family-labour farm households, enhancing their competitiveness in the struggle with capitalist wage-labour enterprises during the depressed agricultural conditions which prevailed from the mid-1870s to 1914, and again during the inter-war years.

At the same time, Rasmussen points out that rising labour productivity was accompanied by falling wheat yields in the period from 1840 to 1880, and productivity gains however measured subsequently 'tended to flatten out, indicating that the first American agricultural revolution was at an end' (Rasmussen 1962: 584).

Before the mid-1930s, in fact, increases in crop yields per acre were a negligible source of output and productivity growth in US agriculture. Mechanization left the vital biological processes which control production and crop yields virtually untouched; indeed, in many areas, its diffusion led to the degradation of the agricultural resource base. Bray and Watkins (1964) emphasize how the mechanization of corn production resulted in 'soil mining' and the rapid depletion of the natural fertility of farm land in the period from 1870 to 1937. The conversion to power farming by tractor and combine in the 1920s also was a major factor in the ecological disaster of the Dust Bowl (Bonnifield 1979). These extreme practices are symptomatic of the secular decline in crop yields experienced in the United States before the 1930s. More fundamentally, however, this decline demonstrated the failure of industrial capitals to appropriate the biological and genetic determinants of agricultural productivity. These factors led the state to redouble its efforts in this area, extending its activities from the collection and dissemination of germplasm to the institutionalization of basic and applied plant genetics research.[1]

The eventual success of this strategy in the US public agricultural research system established the technological base of the modern food system, as we see in the next section.

THE INTER-WAR PERIOD: FOUNDATIONS OF 'FORDIST' AGRICULTURE?

The inter-war years are regarded as a transitional period between extensive and intensive regimes of accumulation, which set the stage for the post-Second World War 'golden age' of fordist accumulation. 'Fordism' as a mode of consumption is based on the expansion of domestic markets for mass-produced products in the advanced industrial economies, even though the internationalization of production is also characteristic of the post-war period. Balanced growth between the capital goods and consumer goods sectors is predicated on the progressive integration of the industrial working class in patterns of mass consumption. This is achieved in an institutional context of Keynesian full employment policies and corporatist politics via rising real wages, extension of the social wage, and an increasing provision of collective goods and services (Glyn *et al.* 1988). Fordist consumption patterns are also characterized by the diffusion of household appliances, including kitchen 'white goods', and convenience foods as industrial capitals appropriate the activities of domestic labour and 'liberate' women for reintegration into the labour force.

These developments have prompted several writers to find rural parallels to industrial 'fordism', particularly in the rise of the corn–soya–livestock complex in the United States and the concomitant diffusion of meat and dairy products as staples of working-class food consumption in the advanced industrial countries (Friedmann 1987; Friedmann and McMichael 1989; Kenney *et al*. 1989). The agro-industrial complex which emerged after 1940 in the United States is seen as the rural complement of the mass-production methods of industrial fordism. That is, the 'packages' of capital-intensive agricultural technologies have ensured the 'mass-production' and declining real prices of grains and livestock products. In the post-war 'golden age' of intensive accumulation, the grains–livestock complex supplied the diet of the 'fordist' working class, underpinning cheap food policies and so contributing to accumulation in the system at large.

A fordist agriculture?

While modern agriculture's contribution to fordist consumption via post-war cheap food policies is clear, the attempt to equate agriculture and industry is misguided. The agricultural labour process remains highly individualistic, entrepreneurial, and provides continuing scope to practice the 'art' of farming. Its industrial counterpart, by contrast, is characterized by a highly complex division of labour, de-skilling, and the destruction of craft knowledge. Capital is notable by its absence from direct production, and especially in the corn–soya complex, and by its concentration in processing and trading. This point is qualified by grower–shipper activities in perishable commodity systems and by contract production in close-confinement intensive livestock systems, but these exceptional cases only heighten the anomaly of grains production. More telling, however, is the argument that agriculture's contribution to fordist consumption and accumulation is based on the *failure* to apply 'fordist' principles to the rural labour process. State intervention and its specific characteristics can be traced directly to this failure of industrial capital to *regiment* and *rationalize* productive capacity. Agriculture remains an unstable, volatile sector of activity, with a pronounced tendency in the absence of state intervention towards low returns and over-production crises.

Agriculture's capacity to supply cheap wage goods did not depend on the diffusion of the new technologies associated with hybrid seeds. The formation of world markets had already created structural over-capacity, as the agricultural depression of the 1920s amply demonstrated. Well before the 'yield explosion' caused by hybrid corn after 1940, agriculture in the advanced industrial countries experienced a virtually perpetual state of crisis. Significantly, Lindert and Williamson (1980) find that falling relative food prices were the main transfer mechanism benefiting low-income urban groups during the 'great levelling' of the US income distribution between 1928 and 1949.

For industrial capitals, the agrarian crisis of the inter-war period was not a supply crisis but a political crisis arising from the impact on agricultural incomes of the trend decline in real prices. The US agrarian populist movements of the late nineteenth century revealed the considerable potential for social mobilization around this issue. State policy intervention in the 1930s was motivated principally by the possible political repercussions of depressed farm incomes rather than by residual fears of urban food shortages. The form of intervention devised in the 1930s in the United States reflected these priorities, although, in supporting farm incomes, state policies exacerbated over-production tendencies.

The real terrain of fordism in the agri-food system is downstream food manufacturing. Here fordist principles are indeed applied to the production process, promoting substitutionist tendencies to reduce agricultural products to interchangeable inputs and to increase value added by conferring industrial qualities, such as convenience and shelf-life, on food. Fordist production organization similarly extends to large-scale catering and particularly to fast food, the fordist food industry *par excellence*, with mass-produced, standardized products and unskilled labour. The giant fast-food empires – McDonald's, Tastee Freeze, Jack-in-the-Box, Burger King, Dunkin' Donuts, Mister Donut, Pizza Hut – were born in the 1950s, accompanying the expansion of those other totems of American fordism, the automobile and suburban housing tracts.

It is at the level of food manufacturing and consumption that the analogy with industrial fordism is most apposite. Agriculture, and particularly the grains sector which is so central to the modern system, continue to resist fordist principles of productive organization. Nature and the culture of 'agri-culture' have yet to succumb to industrial control and process engineering. It was left to the state to tackle the contradictions created but unresolved by agro-industrial capitals, as we see in the following section.

THE NEW AGRICULTURAL TECHNOLOGY/POLICY MODEL

In the 1930s, notably in the United States, economic and political conditions, institutional change and technical innovations led to the articulation of an agricultural technology/policy model which is the basis of the modern agri-food system. The main constituents of this model are the institutional structures of farm support policies which, by regulating commodity markets and preventing marked fluctuations in farm incomes, have reduced production risk and created incentives for farm-level accumulation. With the public agricultural research system disseminating technical innovations and new crop management practices, technological competition became the basis of social reproduction of the farm enterprise and the dynamic behind the sustained productivity growth experienced in the United States since the 1930s. Rural social structures were profoundly transformed by this 'treadmill' of competitive innovation which intensified social differentiation, ownership concentration and the out-migration of the farm population. Concomitantly, this state-subsidized model of farm competition has provided expanding markets for agro-industrial inputs, such as farm equipment, seeds, and agri-chemicals, and ensured low-cost supplies for downstream processing industries and food distributors. The symbiosis between these different but unequal interests, nurtured by economic interdependence and reinforced by political alliances, is a defining characteristic of the modern food system. This capital- and fossil fuel-intensive model emerged initially in the United States, before spreading to Western Europe with post-war reconstruction and Marshall Aid and later to selected sectors of Third World agriculture. In this section, we analyse the origins and diffusion of the agricultural technology/policy model, which is at the root of the transformation and current economic and environmental crisis of modern agriculture.

THE NEW TECHNOLOGICAL BASE

We have seen that industrial appropriation of the energy base and the nitrogen economy had created two powerful sectors of agro-industry by the 1920s. Yet, while these innovations had far-reaching consequences for agriculture and farm structures, the potential synergies between mechanical and chemical appropriations were only realized fully with the diffusion of crop hybridization techniques in the 1930s.[2] These advances in plant genetics overcame one of the fundamental constraints to industrial control of the production process, and brought plant breeding to the forefront of technological change in agriculture. This position was reinforced as chemical and farm equipment companies incorporated genetic research into their strategies of accumulation, promoting greater technological convergence. Plant-breeding programmes could be designed to develop fertilizer-responsive varieties and to adapt plant architecture to withstand mechanical harvesting. As these complementarities were exploited, integrated 'packages' of improved management practices combining mechanical, chemical and genetic innovations were created for the major crop sectors.

Although hybridization so far has proved an elusive prize for leading commercial crops, including wheat, soyabeans, cotton and rice, the experience with hybrid corn established a radically new 'style' of agro-industrial development. Before the introduction of commercial hybrids in the later 1930s, the seed had served a dual function as both product and means of production. With the older open-pollinated varieties, the farmer could keep part of his harvest as seed corn for the next crop without suffering an appreciable loss of yield. For this reason, the genetic information embodied in open-pollinated varieties essentially is a public good (Berlan and Lewontin 1986). However, this alternative effectively was destroyed by the hybrid varieties developed in the public agricultural research system.[3] Thus, 'The F2 generation of hybrid maize, if not biologically sterile, is economically unusable as seed, producing anywhere from 20 per cent to 40 per cent less than the F1 hybrid. *For all practical purposes, such a loss of yield amounts to biological sterility'* (Berlan and Lewontin 1986: 787, emphasis added). One biological constraint was replaced by another, this time giving private breeders potential advantage over farmers. This promise of *de facto* plant breeders' rights and private monopoly rents was significantly advanced by the successful campaign waged by seed firms to restrict public development of commercial hybrid corn lines (Kloppenburg 1984). Under this new institutional division of labour, the private sector soon established a clear lead in corn improve-ment work, giving it the key to control of the production pro-cess (Griliches 1957).

The higher yields obtainable with hybrid seeds had a 'locking-in' effect on farmers as competitive pressures integrated them into crop management systems dictated by agro-industrial capitals. This nexus between genetic innova-tion, typically originating in the public research system, and markets for industrial inputs is graphically illustrated by breeding programmes to develop hybrid varieties adapted to larger plant populations. These are more susceptible to disease and pests and therefore require heavier applications of plant-protection chemicals. In the case of corn, Kloppenburg notes that plant densities doubled between the mid-1930s and the late 1970s, and this crop currently accounts for a third of US herbicide sales and a quarter of the insecticide market. Hybrid corn, he argues, 'has been a major contributor to the historical increase in the intensity of chemical use in American agriculture' (Kloppenburg 1984: 303).

Genetic innovation in the land-grant college research system also was instrumental in accelerating technological convergence in other crop sectors. Cotton, sugar beet and California tomato production provide striking examples of the role of publicly financed research programmes in adapting crops to mech-anization (Rasmussen 1962). The cotton sharecroppers and small farmers of the American South and the *bracero* workers in California were the immediate victims of the ethos of 'farm efficiency' which then dominated public research and extension systems. As Fite observes in the case of cotton, where mechanization research was funded after 1946 by special Congressional allocations,

> There was no way that thirty to fifty acres of cropland could produce enough cotton or grain crops in the Southeast to earn a satisfactory return on the capital investment necessary to fully mechanize production. Tractors, mechanical cottonpickers, and other machines were out of reach of those operators.
>
> (Fite 1980: 203)

In the ensuing exodus, almost a million people left agriculture between 1940 and 1960 in the state of Mississippi alone, a decline in its farm population of 62 per cent (Day 1967). Many other millions of poorer farm families and rural workers also were to lose their livelihoods on the treadmill of technological innovation. These crop 'packages' classically represented by hybrid corn form the technological base of the model of agro-industrial accumulation established in the United States in the 1930s. Many of the characteristic features and contradictions of the modern agrifood system can be traced to the enormous expansion of the productive potential unleased by plant genetic innovations, and the ways in which this capacity has become institutionalized since the 1930s.

PRODUCTIVITY GROWTH AND THE GRAINS–LIVESTOCK COMPLEX

With the main building blocks – tractors and combine harvesters, synthetic nitrogen and hybrid seeds – of the new technology base falling into place in the 1920s, the stage was set to arrest the declining yield trend which had characterized US agriculture in the period from 1870 to 1935. The 'corn yield explosion' led this dramatic reversal as the public research/extension services released high yielding, fertilizer-responsive hybrid varieties adapted to local environments.[4] Corn yields per harvested acre rose from 20.5 bushels in 1930 to 86.3 bushels in 1975, when over 95 per cent of total corn acreage was planted to hybrids (Kloppenburg 1984). The land-saving characteristics of these yield-increasing

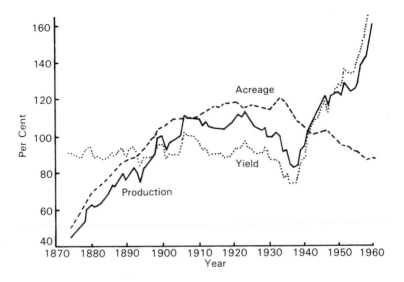

Figure 3.1 US acreage, yield and production of corn 1870–1960. Index
numbers of five-year moving average; 100 = period mean for
each series

Source: James O. Bray and Patricia Watkins, 'Corn production in the United States,
1870–1960', *Journal of Farm Economics* 46 (November 1964): 753

innovations are conveyed by Heady's comment that 'In
1979, corn production of 7.6 billion bushels on 69.4 million
acres would have required 271.9 million acres at the 1910
yield level' (Heady 1982: 31.) These trends are shown in Figure
3.1. The reversal of the long decline and stagnation of US
crop yields per acre from the 1930s undoubtedly reflects
the weight of corn in grain production and demonstration
effects on other crop breeding strategies and farm management systems. The shifting emphasis in American agriculture after 1930 from exploitation of the extensive margin
of productivity growth through mechanical technology to
the intensive margin through biogenetic innovation to raise
crop yields is the true measure of the importance of hybrid
corn. Greater specialization, monocultural regimes and
genetic erosion are the corollaries of this 'method of inventing'.

Innovation of a more conventional, though no less spectacular, kind occurred with the expansion of US soya bean production from the mid-1930s as soya oil conquered markets for edible oils and fats in the food industry and soya meal became established as protein animal feed. This new crop represented the missing link in the grain-fed livestock equation, with corn, and later sorghum, providing carbohydrates and soya meal supplying concentrated protein in compound animal rations. The expansion of soya beans in the United States is a fascinating story of agricultural protection and import substitution, the collection of foreign germplasm, political lobbying and the formation of a powerful agro-industrial 'complex', which has become a central pillar of the modern food system. Industrial compound feeds have replaced farm-produced fodder and feed crops, and livestock production has fallen into the hands of large-scale integrated systems at the expense of diversified mixed farming. As result of these developments, meat products in advanced industrial countries are no longer luxury foods but staples of everyday consumption.

From the 1890s, when the USDA began to introduce new varieties from Europe and the Far East, until the mid-1930s, soya beans were cultivated in the United States principally for hay, pasture or green manure, and only small quantities were processed. The lead here was taken by the Staley Company, one of the latter-day giants of the industry, which established a crushing plant in 1922 to extract oil, mainly for industrial, non-food uses. The market uncertainties which beset farmers and processors, and consequently restricted the cultivation of soya beans as a *cash* crop, were dispelled by the so-called Peoria Plan of 1928. Under this arrangement, an Illinois farmer co-operative agreed to meet a fixed-price supply contract, overcoming processors' fears of idle capacity and unprofitability (Lockeretz 1988). A second major difficulty for the fledgling industry in this early period, when soya beans were just emerging as a cash crop, was the competitive challenge of imported soya oil from China. The American Soybean Association, first established in 1919 and reorganized in 1925, responded by lobbying successfully for tariff protection. Prohibitive duties were imposed on soya beans and oil in 1930, effectively reserving the US market for domestic producers.

This was followed in 1934 by a tax levied on the first-stage processing of coconut oil imported from the Philippines, opening up new outlets for domestic soya oil in the US food industry, which subsequently became its main customer (Berlan *et al.* 1977). Over the next decade, aided by advances in processing, notably hydrogenation, and wartime shortages and price incentives, soya rapidly displaced other vegetable oils as the main source of edible oils and fats.

These efforts to insulate the US market from foreign competition were accompanied by federal-state co-operative research programmes ranging over all aspects of plant improvement, and which maintained close, club-like links with farm organizations, processors, seed companies and users of soya bean products (Lockeretz 1988). These programmes, which developed varieties adapted to local conditions and also to meet specific processing requirements, drew on USDA accessions 'of several thousand varieties from the Orient . . . [which] constituted an enormous gene pool from which specific characteristics could be introduced into the varieties in use in the USA' (Lockeretz 1988: 165).

The expansion of soya bean processing for oil in the 1930s, and particularly during the war, when Asian supplies of palm and coconut oil were cut off, focused attention on ways of utilizing soya meal, the co-product of soya oil. This problem was again tackled by public experiment stations, which devised compound feeds using soya meal for all types of livestock, supported by 'considerable educational and promotional activities by industry and trade groups' (Lockeretz 1988: 159). According to Berlan, this research indicated that concentrated soya protein improved feed conversion ratios, increased livestock growth rates and produced leaner meat which sold at a premium in the market: 'In short, a much more "efficent" system of animal production (in the strictly microeconomic sense) became feasible after the 1930s' (Berlan 1991: 18).

Production of compound animal feeds using soya meal protein concentrates was firmly established during the Second World War; indeed so well that by the 1950s Public Law 480 subsidies and commercial exports were needed to dispose of surpluses of soya beans, oil and meal. The susequent efforts of the American Soybean Association and US policymakers have successfully manœuvered this crop into a commanding position in the modern food system, where it has largely displaced other vegetable oils and protein feed grains. As Lockeretz observes,

> Soybean meal now accounts for about three-quarters of all high-protein livestock feeds, and soybean oil accounts for about three-quarters of all human consumption of fats and oils. By the early 1980s US exports amounted to over 60 per cent of the total consumption of soybean meal and oil in the rest of the world'.
>
> (Lockeretz 1988: 155)

The production potential created by post-war productivity growth also exacerbated the perennial problem of excess capacity inherited from the 1920s and 1930s, depressing commodity prices despite the countervailing effects of farm programmes. As we see below, state intervention in commodity markets, by removing price uncertainty and absorbing surpluses, effectively has institutionalized over-production and created conditions favourable to the sustained growth of output and productivity. Concomitantly, by divorcing farm price policy from income support programmes, the state has found leeway to pursue cheap food policies in the post-war period.

The corn–soya bean–livestock complex was the fulcrum of these cheap food policies, providing the industrial working class and the burgeoning suburban white-collar labour force with a more diversified, animal-protein-based diet. Falling real prices of meat, milk and eggs, reflecting productivity gains in feed crops and new livestock technologies, in turn released purchasing power for the 'white goods' and another mass-produced consumption goods of 'fordist' industry. This interdependence between food policies and industrial recovery is acknowledged in Roosevelt's New Deal pledge to put 'a car in every garage and a chicken in every pot'. This populist phrase also captures the essence of the fordist mode of consumption whose diffusion was the foundation of the post-war 'golden age' of capitalist growth. US agribusiness, and notably the soya bean complex, was both protagonist and beneficiary of the internationalization of animal-protein-consumption norms, leading to what Rama (1985) has described as the 'the international homogenization of wage goods'.

In the United States, the corn yield explosion after 1940 was accompanied by increasing yields of other major grain crops. Between 1945 and 1970, significant average annual yield increases, measured in bushels per acre, were recorded for corn (1.57), wheat (0.56), soya beans (0.35), sorghum (1.41), barley (0.69) and oats (0.51) (Heady 1982). These gains have fuelled a mutually reinforcing pattern of expansion as declining real prices of feed grains provided the stimulus for the reorganization of livestock production. These activities have moved increasingly away from open range and grass-fed operations towards intensive grain-fed systems, epitomized by cattle feedlots and highly mechanized confinement housing for pigs and poultry. This symbiosis has become vitally significant to producers and processors of small grains as the principal outlet of their huge productive potential. The expanding *indirect* consumption of food and feed grains in livestock products is absolutely critical to the workings of the modern food system, and one of its most singular characteristics.

Livestock production is perhaps the one sector of upstream agricultural activity where the parallel with fordist principles of productive organization can properly be drawn. As suggested earlier, it is misleading to extend this to food and feed grains, where biological constraints still obstruct the concentration and centralization of capital. However, these constraints have been weakened significantly in livestock activities since 1950 as industrial capitals have acquired greater knowledge of the reproductive physiology of animals. Fordist production methods have become applicable only as these activities have left the farm.

In terms of fordist organization, the innovations which revolutionized the livestock production process and gave it greater locational mobility are those in genetics, animal nutrition and disease control. These changes, rather than advances in feed crops, are the *sine qua non* of the huge beef cattle feedlots, the drylot dairies, the total confinement units of modern pig farming and battery poultry houses. It is in the automated, controlled environment of the 'factory farm' where the similarities with industrial process engineering and mass production are strongest.[5] These are accentuated by vertical integration with compound feed firms, food

110

manufacturers and distributors and, in some cases, super-market chains.

The new livestock technologies underlying this transition to more concentrated, mass-production operations include genetic innovations in poultry breeding and artificial insemination: 'chickens have been the first to be commercially exploited by the application of inbreeding-hybridization techniques, as earlier used in corn, as well as by methods of selective improvement using the principles of quantitative genetics' (Nordskog 1977 cited in OTA 1981: 170). The breeding stock which produces the 3.7 billion chickens consumed annually in the United States is controlled by less than fifteen primary breeders (1981: 170). The practices associated with the industrial scale and streamlining of modern poultry production have been widely condemned on animal welfare grounds. These practices include reliance on antibiotics, vaccines, growth hormones and other veterinary products to counteract the stresses of total confinement, high densities and restricted space for movement. Modern methods in pig rearing are opposed for similar reasons.

Nevertheless, these developments in animal science have raised growth rates, feed conversion ratios and production densities, with significant gains in economic productivity. For example, average milk yields in the United States increased from 2,000 kg to 4,700 kg between 1945 and 1975, permitting annual milk output to rise by 1 billion kg as the size of the dairy herd fell from 26.6 million cows to 11.6 million (CRS 1984). Similarly, the growth rate of broiler chickens has advanced at 4 per cent annually in recent years, steadily reducing the age at slaughter (OTA 1981). These new livestock technologies, reinforced by higher productivity and production surpluses in feed grains, have led to a spectacular decline in the real price of animal protein, particularly of poultry, stimulating consumption. Thus the retail price of ready-to-cook broilers in the United States fell from 60 cents per lb in 1955–7 to 34 cents in 1976–8, while that of turkeys declined from 66.1 cents to 40.3 cents in the same period[6] (Rogers 1979). Taking a longer perspective, US per capita consumption of poultry meat has risen from around 15 lbs in the mid-1930s to over 70 lbs in the 1980s, when for the first time it exceeded that of beef (Berlan 1991; *The Economist* 21 March 1987). Nevertheless, beef consumption per capita more than doubled between the early 1950s and late 1970s, reducing the share of pork from one-half to one-third in the US supply of red meat. Irrespective of these substitution effects, the pronounced trend to consume grains indirectly in the form of animal protein has changed diets radically in the advanced industrial countries and, increasingly, in the Third World.

In many respects, the development of the modern agri-food system and its internationalization is synonymous with the expansion of the grains–livestock complex and the corresponding diffusion of food consumption patterns based on meat and dairy produce. The insertion of selected Third World economies into the post-war international division of labour in agricultural trade has been heavily conditioned by the accumulation strategies of agro-industrial capitals in this powerful complex, as we shall see in Chapter 4.

The transition to more capital-intensive management systems, involving the substitution of capital and agro-industrial inputs for labour and farm-produced inputs, occurred in the United States from the late 1930s and in Western Europe after 1950. The changes we have described in the preceding sections not only transformed the organization and practice of farming but also brought sustained gains in total factor productivity, that is, output per unit of total input. In other words, although the level of total resource utilization remained virtually unchanged, the process of substitution and capitalization generated rapid productivity growth because the new inputs were more productive. Thus, in the period from 1940 to 1960 when the new agro-industrial model was being firmly consolidated in the United States, total inputs increased at an average annual rate of 0.2 per cent, whereas annual farm output growth averaged 1.9 per cent and crop yields rose by 2.1 per cent. By contrast, the farm labour force declined by 2.9 per cent annually, resulting in spectacular increases in labour productivity (Johnson and Kilby 1975). These post-war trends are placed in a long-term perspective in Figure 3.2.

The post-war acceleration of total productivity growth is viewed from a different angle by Cochrane and Ryan (1981), who suggest that technological change involved the substitution of expensive, relatively inefficient labour by cheap fossil

112

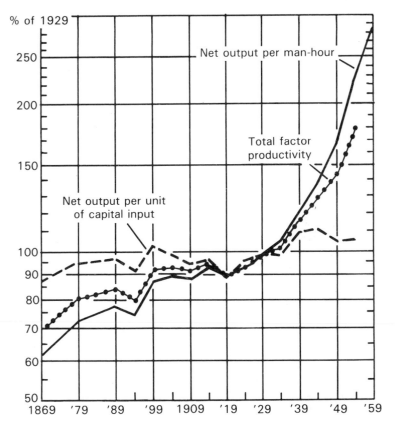

Figure 3.2 Agriculture: labour, capital and total factor productivity
Source: Rasmussen (1962)

fuels consumed in the form of motorized farm equipment, electricity and fertilizers. Although the energy efficiency of agricultural production diminished, the low real price of fossil fuels meant that *economic* efficiency increased.

Thus, the period 1948–73 may be characterized as one of a massive transference of cheap fossil fuels into agriculture in the form of new and improved technologies which had the effect of increasing output per

113

unit of input importantly and increasing total farm output greatly.

(Cochrane and Ryan 1981: 5)

However, this 'production success story' conceals the dramatic restructuring of rural economics and society due to the large-scale expulsion of farm families and agricultural workers, as well as the damage inflicted on the environment by the intensification of production, as we shall see in Chapter 5.

RESHAPING STATE–AGRICULTURE RELATIONS: FROM THE NEW DEAL TO THE URUGUAY ROUND

In this section, we consider the redefinition of state–agriculture relations initiated by the over-production crisis of the inter-war period and cemented by the post-1945 agricultural settlement. The institutional edifice erected in these years represents the compromise constructed by the state in pursuing its cheap food strategy while retaining political legitimacy with the disparate interests in the 'farm bloc'. This political constraint led to wholesale intervention in heavily protected markets to counter the depressive effect of rapid technological advance on farm incomes and rural employment. The structural tendencies unleashed by this historic compromise subsequently found international expression in the instability of agricultural commodity markets in the 1970s and the 'farm crisis' of the 1980s. This recent crisis has provoked fundamental questioning of the high-input, 'productivist' farm policy model inherited from the New Deal period, which is reshaping the policy agenda for the 1990s.

Following the appropriationist framework outlined earlier (pp. 90–1), we would argue that the specific characteristics of state intervention were strongly influenced by the technical and social organization of the production process. This is characterized mainly by many atomistic units producing standardized commodities and lacking market power. As we have seen, the formation of a world market resolved the 'agrarian question' as disputed class hegemony, although food surplus mobilization and provisioning objectives continued to be state priorities. With the onset of the Great Depression in

114

the later nineteenth century, the social dimensions of the 'farm problem' became increasingly prominent, with millions of family producers earning relatively low returns, often barely at subsistence level. In these circumstances, to attempt an unrestrained competitive solution to the existence of excess capacity would have extremely damaging social and political consequences.

In advanced capitalist societies, which eschew collective forms of productive organization and state ownership of the land, the anomaly of agriculture is that, due to the weakness of concentration and centralization tendencies, competitive regulation remains the *only* alternative to state intervention. Since industrial capitals are outside the immediate production process, the 'rationalization' of excess productive capacity must be managed by state policies if the downward trend of agricultural prices and farm incomes is to be halted.[7] In the 1930s, as a consensus on farm support policies was emerging, hybridization opened up new sources of productivity growth based on highly intensive production systems. State intervention therefore became even more indispensable to arrest the long-term decline of farm incomes and calibrate the inevitable exodus from agriculture.

Yet at that time advanced capitalist economies were only just beginning to face up to the consequences of the chronic excess capacity left by the previous phase of agricultural expansion. Initially, as wartime programmes and guaranteed prices were abolished, this adjustment was left to market forces, provoking the world-wide collapse of commodity prices. These conditions have prompted Kindleberger (1987) to observe that 'it is an open question whether an independent depression in agriculture helped to cause the stock market crash' of 1929 and the crisis of extensive accumulation.

The institutional framework which emerged from this crisis in the 1930s reflected radical changes in relations between the state and agriculture, most notably in the United States. At the macro-economic level, New Deal farm programmes attempted to reconcile cheap food policies with a political commitment to maintain farm incomes. This commitment was expressed in an explicit concern for rural welfare, complementing immediate goals to stimulate the recovery of consumer markets. The new policy framework was meant to alleviate

115

the depressive impact of over-production on farm incomes but the dynamic growth of productivity continually increased productive capacity and the potential size of surpluses. The result resembled an unending game of leapfrog as rising yields constantly negated the contractionary effects of acreage allotments and land retirement schemes.[8] In these circumstances, when minimum price levels were adopted as the preferred mechanism of farm income support, as in the case of the US New Deal farm policy, the effect was to institutionalize, not reduce, excess capacity. Programme costs escalated sharply, progressively involving the state in virtually all aspects of agricultural production and marketing.

Direct, detailed intervention in agricultural supply was used not only to maintain cheap food policies, including extension of the social wage via food assistance programmes, but also increasingly as a weapon of Cold War politics and to open up export markets in the Third World.

The failure to restrict over-production by counteracting productivity growth put constant downward pressure on commodity prices, and hence farm incomes. With rising input costs, farmers were caught in a cost–price squeeze which, in turn, encouraged rapid innovation. As price-takers in regulated markets, farmers had every incentive to reduce unit costs and maximize output, contributing to the magnitude of the underlying problem of excess capacity and surplus stocks. It is this powerful combination of cost–price squeeze and competitive innovation which changed the face of agriculture and rural society after the Second World War.

Farm policies adopted in the New Deal era set the pattern of state–agriculture relations for the next fifty years, in both the United States and Western Europe. These policies were legitimated by the state's strategic concern for food supply and its political accommodation with farm organizations and agro-industrial interests.

New Deal farm policy in the United States

Despite increasing political mobilization by farmers in the 1920s, when President Coolidge twice vetoed legislation to relieve the post-war depression in agriculture,[9] farm policy only moved on to the agenda of national party politics in the

1930s. Finegold (1985) suggests that a necessary condition for this shift was the decision by the Eastern business establishment to moderate its opposition to state intervention in agriculture in the interests of wider economic recovery.[10] The sufficient condition for this to occur, Finegold argues, came with Roosevelt's election in 1932, achieved by combining the Democrat's traditional ethno-cultural politics with a policy-based alliance with the leading farm associations.

The cornerstone of New Deal farm policy was formed by the Agricultural Adjustment Acts of 1933 and 1938, which aimed to raise farm incomes by restoring 'equality of purchasing power' for agricultural products, taking 1910–14 relative prices as parity. Producers who contracted with the Agricultural Adjustment Administration (AAA) to restrict their acreage under cultivation were then eligible for federal loans from the Commodity Credit Corporation (CCC) on commodities placed in storage. The loan rate effectively became the minimum support price since, if market prices fell below this level, farmers could sell their crops into CCC stocks at the agreed rate. Compliance with these production controls at the local level, including the distribution of acreage allotments, was supervised by farm organizations. The AAA's programme originally embraced seven basic commodities – wheat, corn, cotton, rice, tobacco, hogs and dairy products – and the CCC's loan rate at first applied only to wheat and cotton. However, both programmes were soon extended to other products, broadening farm support for state market intervention and establishing 'a powerful coalition of special farm interests' in Washington (Fite 1981: 57). Farm organizations, in effect, became an integral part of the institutional apparatus of the state.

The central activities of the AAA and CCC in production control, acreage allotment, commodity storage and minimum price programmes were accompanied by a broad array of new farm policy initiatives. These included emergency farm mortgage relief, the expansion of federal farm credit programmes, soil conservation, surplus food distribution, and support for rural electrification. By the late 1930s, 'Under pressure from farmers, Congress and the President had injected the federal government into agriculture to an extent thought both undesirable and impossible only a decade earlier' (Fite 1981: 64).

Nevertheless, since acreage restriction was the principal method of production control, the main beneficiaries inevitably were larger commercial farmers who could afford to take land out of production: 'No part of the AAA had anything significant to offer to the hundreds of thousands of sharecroppers, tenants, and those on small acreages in such poor farming areas as Appalachia' (Fite 1981: 61). Fite estimates that there were barely three million viable commercial operators at the end of the 1930s out of a total of 6.5 million farmers. Hadwiger also recognizes the large farm bias that was built into the provisions of the AAA, observing in the case of wheat commodity programmes that 'Program rules have been framed for the convenience of the larger farmer. The acreage cuts have been inherently unacceptable to the farmer whose total farm size was already inadequate' (Hadwiger 1970: 317).[11]

These New Deal programmes established the institutional framework of the US technology/policy model. This framework was consolidated during the Second World War and the post-war period as state intervention to support farm prices and incomes became enshrined in the corporatist politics of the 1950s and 1960s. If distributional issues are set aside, the New Deal programmes improved farm conditions and net farm incomes rose substantially between 1933 and 1940. Nevertheless, prices averaged 20 per cent below parity and annual per capita income of the farm population was just over a third of that received by the non-farm population (Fite 1981). Fite's assessment of farm policy at the end of the New Deal era is just as apposite for the early 1990s:

> The problem centred around continued heavy production and price-depressing surpluses, despite acreage restrictions . . . A crisis of abundance in agriculture was at hand by 1940 . . . Some major intervention by government would have been required to save the situation from disaster if the demands of World War II had not wiped out the huge supplies.
>
> (Fite 1981: 65)

These comments reveal the difficulties of intervention in a sector where production capacity is determined by the interaction of scientific innovation, uncontrolled biological factors and the decisions of millions of individual producers. Moreover, in 1940, the engine of productivity growth, fuelled by the 'packages' of genetic-chemical innovations being disseminated by the public agricultural research and extension system, had barely started to gain momentum.

Western Europe

Once the food shortages of the immediate post-First World War years disappeared, farmers in Western Europe experienced similar difficulties to their US counterparts as prices declined and remained at depressed levels. This situation was exacerbated by the dismantling of wartime production incentive schemes, and British farmers suffered the 'great betrayal' of August 1921 when legislation to continue guaranteed prices indefinitely was revoked. During the 1920s, most Western European countries, apart from Britain, gradually reimposed agricultural tariffs and levels of protection rose sharply in 1929–31 as the crisis deepened. Subsequently, tariffs were complemented by more direct controls, including import quotas, import licence fees and 'linked-utilization' measures intended to ensure the use of home-produced supplies (Tracy 1982). Import controls were usually accompanied by direct intervention in the organization of domestic agricultural markets, as in the case of France, where a government agency, the *Office du Blé*, created in 1936, was empowered to set the price of wheat and purchase surpluses. Although methods and schemes might differ in their details, virtually all Western European countries resorted to some form of state intervention in agriculture to control production, organize markets and restrict imports. The fascist regimes in Germany and Italy carried this national policy response to the crisis to extremes, instituting central planning to achieve the goal of national self-sufficiency in food.

Britain also followed this general pattern of response, abandoning agricultural free trade in 1931 to halt the dumping of world surplus exports and the collapse of domestic farm prices. However, agricultural protection was in conflict with imperial preference and the maintenance of overseas markets for industrial exports. These priorities were reflected in the series of trade agreements that Britain negotiated with overseas primary producers, both imperial and foreign, in the 1930s. Thus Rooth contends that 'the terms of the treaties dictated the eventual shape of agricultural policy in the UK, helping to determine the emergence of Exchequer-financed subsidies as the principal support device between the late 1930s and the 1970s' (Rooth 1985: 174).

Experiments with alternative support policies, such as import quotas and a levy–subsidy system, were short-lived and foundered on the concessions made to primary producers in the Dominions and other traditional food-exporting countries. 'The ultimate form of protection thus was shaped by the agreements. They helped to ensure that low prices prevailed, that consumption was encouraged, and that British farmers were maintained in a way that inflicted minimum damage to exports' (Rooth 1985: 190). This protection was applied piecemeal in the 1930s on a commodity-by-commodity basis using different policy measures and in various combinations, including price subsidies or 'deficiency payments' for wheat, subsidies and import restrictions for beef, monopolistic Milk Marketing Boards, and marketing schemes and import controls for other crops (Tracy 1982).

The post-war settlement for
Western European agriculture

The similarity of national policy responses in Western Europe to the crisis of the 1930s also can be observed in the Second World War and the period of post-war reconstruction. Strategic considerations, food shortages and the scarcity of foreign exchange forged a common emphasis on the maximization of food production and involved the state in all sectors of the agri-food system. In Britain, farmers 'had a good war' and average farm real net incomes more than tripled (Bowers 1985). Moreover, this return to agricultural prosperity brought higher social status and renewed political influence, which were used to good effect in peacetime. Farmers in continental Western Europe were much less fortunate but output recovered rapidly from the ravages of war and the lack of industrial inputs, assisted after 1948 by Marshall Plan programmes. Pre-war output levels were regained by the late 1940s and surpluses began to reappear in the mid-1950s.

Nevertheless, the harrowing wartime experiences of food shortages, rationing, and dependence on imported supplies left an indelible stamp on the post-war settlement for agriculture in the major West European countries. In Britain, the nature of this settlement already was discernible in the early 1940s as the state pledged to maintain fixed prices and guaranteed markets for the duration of the war and in the first year of peace. These wartime policies and priorites became the basis of the 1947 Agricultural Act, which set the policy agenda for the next twenty-five years in Britain. In addition to guaranteed prices for major products, which were implemented by 'deficiency payments' financed by general taxation following the removal of wartime import controls, the 1947 legislation gave the farmers' unions the statutory right to consultation in the Annual Review of prices and agricultural conditions (Bowers 1985). Leaving aside the British preference for deficiency payments rather than import levies as the method of farm support, the watchwords of the post-war settlement in Western Europe were protection, output expansion and technical efficiency. Farmers' organizations also pressed with varying success for statutory requirements to maintain parity between farm and non-farm incomes but this issue frequently became confused with structural policy and other goals.[12]

The post-war settlement was implemented using the familiar instruments of guaranteed prices, state regulation of major commodity markets, and their insulation from foreign competition. The stimulus to output growth from these protective trade policies, or, in the British case, deficiency payments, was reinforced by efforts to accelerate innovation and promote technical efficiency. Thus the state embarked on modernization programmes involving investment grants, input subsidies, special credit and fiscal incentives, public agricultural research and its dissemination by extension services. In France, for example, the first of several Modernization and Equipment Plans was adopted for the years 1947–50 (Tracy 1982). Farmers in Britain received payment of direct grants for capital investment, chemical inputs and the adoption of new techniques. Moreover, the share of this direct grant element in agricultural support rose considerably in the 1950s and 1960s. This change in the relative weight of price support and capital and input subsidies, reinforced by a policy of maintaining guaranteed

121

price increases below cost increases, meant that higher net farm income effectively could be obtained only by innovation and productivity growth (Bowers 1985).

As the problems of post-war reconstruction receded and gave way to surpluses in the mid-1950s, the policy emphasis shifted from across-the-board growth towards 'selective expansion'. However, these attempts to restrain output, even of commodities in surplus, were quickly undermined by the accelerated growth of productivity and the opposition of the agricultural lobby. In Britain, this call for selective expansion was merely a temporary interlude and, once it had been decided in the early 1960s to limit the Exchequer's contribution to the agricultural support system by shifting part of the burden on to the consumer via import restrictions, expansion again became the order of the day (Bowers and Cheshire 1983). As the European Community of the Six moved towards agreement on the Common Agricultural Policy in the mid-1960s, state–agriculture relations presented an intricate web of protection, support systems, rising surpluses and formidably strong farmers' organizations.

The United States: wartime prosperity and post-war surpluses

The Second World War ushered in twelve years of virtually uninterrupted agricultural prosperity for American farmers but, without such foresight, farm organizations concentrated on achieving the long-cherished ambition of parity price levels. These demands were accommodated within the framework of New Deal programmes and institutions now shorn of their reformist zeal. As in the United Kingdom, American farmers were anxious to avoid a repetition of the events of 1920–1, when their wartime prosperity was quickly eroded by the collapse of farm prices, and this influenced their response to calls for all-out production and demands for emergency price control legislation. Under legislation enacted in 1942, farm price ceilings were restricted to 100 per cent of parity but the farm bloc successfully supported the Steagall Amendment, which provided guaranteed price supports at 90 per cent of parity for two years after the end of the war. In the event, prices generally remained above the support levels due to the extraordinary needs created by the war and peacetime reconstruction overseas.

With these policy assurances and buoyant demand, wartime crop and livestock production reached record levels and net farm income rose from US$4.5 billion to $12.3 billion between 1940 and 1945. Rising productivity growth was the main source of these impressive wartime gains, marking the wider dissemination of mechanical and chemical inputs. In many respects, the war years were a period of transition to the new capital-intensive agro-industrial model, encouraged by firmer prices and labour shortages. The increasing capitalization of US agriculutre is reflected in the expansion of gross farm output, which grew by 14 per cent in the years 1940–5, whereas acreage rose by only 5 per cent and the farm population fell by 5.1 million due to the demands of military service and the recovery of urban-industrial employment opportunities (Fite 1981).

In the 1950s, US farm politics were dominated by rising farm surpluses and their financing and declining relative prices, which tightened the cost–price squeeze on farm incomes. With output expansion exceeding the slow growth of aggregate demand for food by a significant margin, farm prices and average net farm incomes were under heavy pressure for most of this decade.[13] As innovation and productivity growth continued, opposition to the New Deal policy of high and fixed price supports quickly gained ground, although the Korean War brought temporary respite. After bitter controversy, the Eisenhower Administration finally prevailed and flexible price supports were introduced by the Agricultural Act of 1954 in an attempt to allow greater rein to market forces.

This legislation was followed by two programmes which mounted a direct assault on the surplus problem. The 'Food for Peace' (Public Law (PL) 480) programme of July 1954, was used to dispose of surpluses, especially of wheat, through non-commercial channels by increasing food aid to developing countries. As we shall see in Chapter 4, this programme was of major importance in the disarticulation of staple food production in these countries and in the creation of export markets by changing food consumption habits towards US patterns. Thus Cochrane and Ryan suggest that PL480 'may well have saved farmers and their programs from complete disaster in the late 1950s' (Cochrane and Ryan 1981: 77). The second initiative was the Soil Bank programme of 1956, which attempted to reduce production capacity directly rather than alleviate its consequences. This programme introduced fixed cash payments for farmers who contracted to retire land either for three years in an acreage reserve or for a longer period in a conservation reserve. The first component of the Soil Bank programme was abandoned in 1958 but the conservation reserve continued until 1972; at its peak in the early 1960s, the acreage in the programme exceeded 25 million acres, mainly land previously planted to feed grains and hay (Cochrane and Ryan 1981).

Despite these new initiatives, the issue of excess production and the high costs of farm programmes remained acute.

> The 1950s ended with surpluses, distressed farm incomes, large-scale non-commercial government exports, and sky rocketing productivity advances. . . The decline in income occurred despite massive removal of farm products from commercial markets. The value of government-owned stocks increased to over $6 billion and shipments under P.L.480 were valued at more than $1 billion in each year since 1956.
>
> (Cochrane and Ryan 1981: 35)

The escalating costs of farm programmes, aggravated by popular reaction against 'paying farmers to do nothing', became a prominent political issue in the early 1960s. Attention under the Kennedy administration focused on 'supply management' proposals, mandatory versus voluntary production controls and the pros and cons of 'getting government out of agriculture'. The heat eventually was taken out of these debates when wheat producers voted against mandatory controls in the wheat referendum of May 1963. Both the Johnson and Nixon administrations endorsed the principles of voluntary land retirement programmes but sought to bring support prices closer to world market levels by relying more on direct payments ('deficiency payments' in UK parlance) to maintain the incomes of producers participating in such schemes. However, even voluntary land idling measures implemented under these supply adjustment programmes

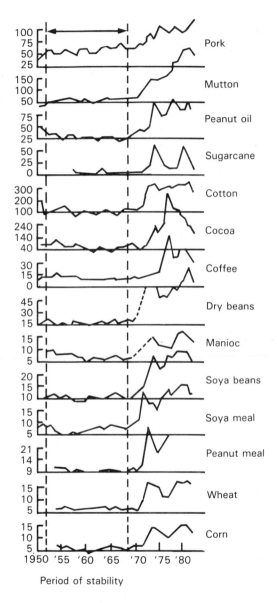

Period of stability

Figure 3.3 World price trends of major agricultural commodities (average annual prices in cents US$)

Source: M. Marloie, *L'internationalisation de l'agriculture française*, (Paris, Editions Ouvrieres, 1984, p.14). Calculated from selected price series of the FAO

were soon to appear unnecessarily restrictive in the increasingly unstable world economic conditions of the the early 1970s.

Instability and farm crisis

The collapse of the Bretton Woods system and the accompanying boom in world output led to an extraordinary upsurge in commodity prices, which suddenly appeared to offer a market solution to the US farm problem (Figure 3.3). Wheat prices rose from an annual average of US$1.58 per bushel in 1971 to $3.81 in 1973 and $4.90 in 1974, while net farm income more than doubled between 1971 and 1973, from US$14.6 billion to $33.3 billion. These events, reinforced by devaluation of the dollar, persuaded the Nixon Administration to follow more market-oriented farm policies centred on a strategy of export expansion. This strategy became the main focus of efforts to eliminate excess capacity and so reduce the role of costly price supports and production controls in maintaining farm incomes. The spectacular rise in international commodity prices which followed the 1972 US–Russian grain deal not only gave credence to this strategy but convinced many of the imminence of world food shortages and famine. The US embargo on soya bean exports announced in June, 1973, and the doubling of soya bean prices which followed, seemed to confirm the new climate of scarcity.

These developments brought to an end the period of unprecedented stability in world grain markets founded on US production controls, stockpiling, subsidized exports and PL480 food aid. Soaring grain prices, falling world stocks and domestic food inflation in 1973–4 led to the withdrawal of US acreage restrictions in 1974, although these and other traditional farm support mechanisms were reinstated in 1977 as fears of a world food crisis receded. Moreover, by the time of the World Food Conference in 1974, the importance of PL480 funds in financing US agricultural exports had fallen to 4 per cent and programme allocations were at their lowest level in two decades. When Third World countries were struggling under the exchange constraints imposed by the quadrupling of oil prices in 1973–4, PL480 food aid was concentrated for 'political' reasons in South-east Asia (Destler 1978). The pillars supporting the international food order, epitomized by low and stable world grain prices, thus collapsed in the mid-1970s and US leadership in agricultural trade, as in other sectors, came under increasing challenge.

The United States was not alone in looking to export markets to resolve the contradictions of its domestic farm policies and reduce escalating support costs. The US technology/policy model was transplanted successfully to Western Europe in the post-war reconstruction when cheap food policies were pre-eminent. Productivist strategies subsequently were fostered within the highly protective framework of the Common Agricultural Policy (CAP), which underwrote the subsidies paid to produce, store and export the mounting food surpluses. These costs have absorbed a progressively rising share of EC expenditures, enhancing the priority given to agricultural export promotion and bringing the EC increasingly into conflict with the United States. Until the 1980s, broadly speaking, this conflict was muted due to the world food shortages of the mid-1970s, continued debt-financed purchases by Third World countries and increasing sales to capital-surplus OPEC members. US agricultural exports tripled from US$7.2 billion to $21.8 billion between 1970 and 1975, and then doubled again to a record level of $43.8 billion in 1981, when the sharp appreciation of the dollar, world recession and the Third World debt crisis led to a marked fall in US export value, volume and market share.

With roughly one acre in three by then used for export production, this decline aggravated excess capacity, undermined farm income and raised US farm support costs significantly. The reversal of export expansion was a major contingent cause of the farm crisis which gripped US agriculture in the early and mid-1980s. During the 1970s, the exceptional conjuncture of rising export sales, favourable world market prices, and low real interest rates, encouraged high levels of debt-financed investment in agricultural assets. After 1979–80, however, producers following this strategy of business growth were caught in a classic debt trap of sharply rising real interest rates, falling farm incomes and collapsing land values. Highly-leveraged producers were the main long-term casualties of the US farm crisis, but most of the producers suffered varying degrees of 'financial stress' in the 1980s (Buttel 1989a).This crisis was the hardest recession to hit US agriculture since the 1930s and the farm population declined by about a sixth, or 1 million, between 1980 and 1987. Despite massive state support, median farm income fell to 79 per cent of the non-farm family median, and non-farm sources accounted for 46 per cent of farmers' earnings (USDA 1989). EEC farmers experienced broadly similar conditions of declining farm incomes and asset values, although the incidence of acute farm indebtedness was less marked.

The slower growth of agricultural trade in the 1980s, together with the deteriorating domestic farm situation, has intensified competition between the United States and the European Community for export markets, vying with each other to match price subsidies and sales conditions. These aggressive trade policies, exemplified by the US export enhancement programme and EC export restitution payments, have provoked increasingly acrimonious disputes, which have verged on trade war on several occasions. These include the so-called 'pasta wars', the continuing row over growth hormones in US meat exports and US objections to the loss of export markets following the accession of Spain and Portugal to the European Community. The heavily subsidized competition between the two agricultural 'super-powers' inevitably has drawn specialized 'third' country agricultural exporters into the fray, threatening their export earnings, farm incomes and, in such cases as Argentina and Brazil, their capacity to service their debt. This market instability and their seriously disadvantaged competitive position led to the formation of the fourteen-strong Cairns Club group of 'third' countries to give weight to their demands for the reform of world agricultural trade policy.

Farm policy reform and new constituencies

This call for freer world commodity markets and the abolition of protectionist farm support structures has been echoed by international organizations, such as the OECD and World Bank, and was affirmed at the 1987 Venice Summit. In July 1987, this orientation became official US policy in the Uruguay Round of GATT negotiations on farm trade. The US presented sweeping proposals to phase out export subsidies, import restrictions and all forms of farm support linked to production by the year 2000, the so-called 'zero option'. The United States moderated this uncompromising stance in subsequent negotiations but it remains an intransigent opponent of EC export subsidies. The European Community espouses a more gradualist approach to dismantling farm subsidies, although the imposition of milk quotas in 1984 and the system of agricultural expenditure stabilizers and set-aside scheme agreed at the emergency EC summit in Brussels in February 1988 together represent an historic turning point in European farm policy. The failure to reconcile the sharp divisions on farm trade between the United States and the Cairns Group, on the one hand, and the European Community and Japan, on the other, provoked the breakdown of the Uruguay Round in December 1990. The GATT Secretariat is currently seeking to reopen discussions but there are few signs at the time of writing that these efforts will quickly produce agreement on a radical timetable of agricultural reform and movement towards free trade.

These events reveal that state–agriculture relations are now under great strain as pressure grows to break away from the corporatist arrangements established by the post-war settlement. This mould has been cast aside in other sectors and it remains to be seen whether farming will continue to receive special treatment by virtue of its ideological and cultural resonance in an urban world. Certainly the neo-liberal critique of state intervention, which has become influential in most Western economies, whether under social democratic or conservative governments, is equally applicable to the subsidies paid to farmers. However, for the moment, Western governments display indecision, hesitating between free market policies and greater reliance on administrative regulation and *dirigiste* structural programmes to resolve the complex, multi-faceted problems of modern agriculture and rural society.

This indecision also reflects the greater political significance of farm policy, which increasingly is pervaded by 'green issues', ranging from pollution to the kind of rural civilization which should be maintained.[14] In Europe, the campaign to control CAP expenditures led by the Thatcher government heightened public awareness of the grain 'mountains' and wine 'lakes', as well as their cost in terms of higher nitrate and pesticide levels in the environment. Public disquiet over such profligate abundance is deepened by the harrowing pictures of hunger relayed by Western media from outlying areas of the 'global village'. The proliferation of environmental movements, pressure groups and the electoral campaigns waged by Green parties in Western Europe have created a well-informed articulate opposition against the intensive agricultural practices which produce these surpluses. These new movements have demonstrated their ability to mobilize public opinion on a wide variety of farm-related issues, such as water pollution, animal welfare and the destruction of wildlife habitats. Consumers and taxpayers increasingly recognize that it is their funds which underwrite these environmentally harmful activities, whose costs also are reflected in inflated food prices (National Consumer Council 1988). Food quality and the way it is produced, including the use of pesticides, growth hormones and high-density animal housing, have aroused concern among consumers and rein-forced the environmental critique of intensive farming methods, generating demands for organic produce and more humane livestock rearing.

The farm crisis of the 1980s has taken many guises – over-production, intensification, rural depopulation and poverty, the fiscal strains of agricultural protection, trade reform and environmental problems - and its legacy is an extremely complex agenda for farm policy change, which must now accommodate a much wider, more pluralistic constituency of interests. Farm and food politics and the future directions of the agri-food system no longer are the preserve of the narrow, club-like 'policy community' formed by representatives of the 'farming industry', agribusiness, the agricultural research establishment, regulatory quangos and ministry mandarins. Consumers of both food and the rural environment where it is produced will have a powerful voice in the redefinition of state–agriculture relations in the 1990s.

NOTES

1 The task of collecting seeds and crop varieties was organized in 1898, when the United States Department of Agriculture established the Office of Seed and Plant Introduction.

2 On the growth of Mendelian-chromosome genetics theory in the early twentieth century, which supplied the theoretical foundations for the new plant-breeding techniques, see Allen (1978; 1979).

3 For further discussion of the development and diffusion of hybrid corn, see Griliches (1957), Hayami and Rutten (1971), Bogue (1983), and Kloppenburg (1984, 1988a).

4 Griliches observes that 'hybrid corn was a method of inventing, a method of breeding superior corn for specific localities. It was not a single invention immediately adoptable everywhere. The actual breeding of adaptable hybrids had to be done separately for each area' (Griliches 1957: 502).

5 Even so, these parallels are significantly weaker in beef production as commercial feedlots essentially fatten and 'finish' animals supplied by independent cattle ranchers. Beef production is further distinguished from poultry operations by the continuing competitiveness of range-fed methods in supplying large sectors of the market, such as the fast-food industry. By contrast, range-fed poultry products are sold mainly in 'niche' markets created by ethical values and nutritional concerns arising from the use of growth hormones and other pharmaceuticals.

6 These prices are deflated by the consumer price index with 1963–7 as the base line.

7 The term rationalization is chosen deliberately given the emphasis in official rhetoric after the Second World War on 'farm efficiency'.

8 Oppostion to the introduction of mandatory supply management programmes and direct income payments is discussed by Cochrane and Ryan (1981).

9 The presidential veto was used to reject two versions of the McNary–Haugen bill, which proposed to raise domestic farm prices by buying up surpluses for sale abroad.

10 One-quarter of the US labour force was still employed in agriculture in 1929.

11 The selectivity of New Deal farm policy was accentuated in the late 1930s when attempts to formulate broader social programmes through the AAA, which would address rural poverty problems more directly, were defeated. This culminated in the abolition in 1946 of the Farm Security Administration (FSA) and its programme of farm-ownership loans to sharecroppers, tenants and farm workers.

12 Attempts to address the parity issue include the 1955 Agricultural Act in West Germany and the Loi d'Orientation of 1960 in France (Tracy 1982).

13 Cochrane and Ryan (1981) cite research results which suggest that farm prices in the 1950s would have been 10 to 25 per cent lower in the absence of price support programmes. Excess capacity is estimated to have averaged between 5 and 9 per cent of annual output in this decade.

14 In a speech at Chatham House in January 1987, President Mitterrand opposed a 'purely industrial' form of agriculture with large agri-

food factories and no 'peasants and farmers'. The founders of the Community, he argued, had envisaged 'a certain kind of rural civilization' for Europe, which had to be taken into account as well as profitability (see also p. 66).

4

INTERNATIONALIZATION AND THE THIRD WORLD FOOD CRISIS

In order to grasp fully the implications of the US technology/policy model, we need to examine the circumstances under which this domestic model was converted into a global system. This chapter addresses the contradictions between state-supported productivist agriculture in the advanced industrial countries and the continued satisfaction of minimum food requirements in the Third World. Initially, however, we consider the diffusion of the US model in Western Europe and the emergence of the post-war international food order.

Agriculture in the advanced industrial countries inherited an enormous overhang of excess productive capacity from the First World War which depressed farm prices and incomes in the 1920s and *a fortiori* in the 1930s. In the United States market forces made some inroads into this problem, but the reformist politics of the New Deal ruled this out as an acceptable solution. Rather, as we saw in Chapter 3, the institutional changes in US agricultural policy from 1933 to 1938 greatly extended state intervention in commodity markets to support farm incomes by idling productive land, stockpiling production surpluses and establishing minimum price levels. These institutional reforms significantly reduced the risk and uncertainty caused by market instability, creating a favourable climate for techno-logical innovation. Further encouragement came from the reorganization of farm credit programmes, which increased the supply of relatively low-cost capital, the introduction of comprehensive crop insurance in 1938 and the strengthening of the public agricultural research and extension systems (Farrell and Runge 1983). These reforms provided the institutional foundations for the era of 'science power' (Lu *et al*. 1979) in

American agriculture, which saw total factor productivity increase by nearly 150 per cent by the mid-1970s. With the research/extension complex forcing the pace of innovation and diffusion, American farmers were drawn onto a 'treadmill' of increasing technological competition, leading to the 'chronically maladjusted farm sector' which characterized the 1950s and 1960s (Cochrane and Ryan 1981).

Against the background of the Cold War and the United States' unrivalled leadership in the world economy, dissemination of the US agro-industrial model to Western Europe, Japan and later to selected Third World countries established a new international food order under American hegemony (Friedmann 1982). This supremacy, maintained throughout the 1950s and 1960s, the 'golden age' of intensive accumulation, was distinguished by the remarkable stability of world commodity markets. These stable conditions were maintained by US acreage restrictions and storage programmes and the protectionist orientation of US agricultural trade policies. Before 1962, when domestic support prices were lowered to world levels, overseas markets were important primarily as an outlet for crop surpluses held by the government rather than as a direct source of foreign exchange and farm income (Revel and Riboud 1986). Until the Kennedy years, therefore, the US exerted its dominance to achieve geo-political advantage and commercial position via subsidized exports and food aid, which secured the orderly disposal of surplus stocks and greater market penetration for US agribusiness.

Reflecting agriculture's role in the national economy, the international food order maintained low prices of basic food and feed grains and other staple foodstuffs, complementing the wider framework of international economic regulation instituted by the Bretton Woods agreement. The economic policy discipline imposed by this regime certainly was made less onerous by the stability of commodity prices (Maddison 1982). Although its technical and social organization resisted fordist principles, US agriculture made a substantial contribution to accumulation in the post-war fordist era, both domestically and internationally.

In the Third World, US crop surpluses, initially in the guise of food aid, have been instrumental in creating food import dependence, reversing pre-war patterns of agricultural trade.

Subsidized US food exports have undermined peasant agriculture and rural self-provisioning, integrating production–consumption more firmly into the circuits of agro-industrial accumulation. The result has been to internationalize the resource base of agriculture, which is reflected in the increasing asymmetry between rural supply structures and domestic food requirements. As in the transition to intensive accumulation in the advanced countries, the incorporation of Third World countries into the post-war international food order has destroyed traditional ways of life by extending commodity relations to the food supply. Friedmann argues persuasively that this order 'occupied a specific place in the dynamics of global accumulation', representing 'an intrinsic part of the project of capitalist industrialization' and contributing 'to the widening and deepening of capitalist relations within the world economy' (Friedmann 1982: 255–6). This chapter examines the internationalization of the US technology/policy model after the Second World War, and particularly its consequences for agrarian structures and food supply in the Third World.

INTERNATIONALIZATION OF THE US AGRI-FOOD MODEL

Diffusion of the US agri-food model occurred within the international framework of fordist regulation established after the Second World War, and epitomized by the commitment to private enterprise, free trade and multilateralism as mechanisms to consolidate American hegemony. The new international economic order was articulated through a 'coherent set' of international monetary, financial and trading arrangements, whose institutional expressions were the International Monetary Fund (IMF), the International Bank for Reconstruction and Development (IBRD) and the General Agreement on Tariffs and Trade (GATT) (Glyn et al. 1988). Establishment of a liberal economic order to support American geo-political objectives gained greater urgency with the onset of the Cold War, which led in 1947 to the launching of the European Recovery Programme, more widely known as the Marshall Plan. Grants and donations of US agricultural commodities played a prominent part in relieving immediate post-war food shortages in Western Europe and shipments increased under the Marshall Plan, securing an important bridgehead for US exports. According to Friedmann, 'Marshall Plan aid in the form of food, feed, and fertilizers . . . was geared both to providing immediate consumption needs during economic reconstruction and to rebuilding agriculture' (Friedmann 1982: 261).

European agriculture rapidly regained pre-war production levels and, with the reappearance of commodity surpluses in the early 1950s, the focus of policy shifted towards more selective expansion, with modernization and technological efficiency as the watchwords. The post-war transition to capital-intensive agricultural systems in Western Europe proceeded at different speeds but, as Tubiana observes,

> the essentials never changed: the capitalisation of farms (mechanisation) and a massive reduction in the agricultural workforce; specialisation according to locality and within the production process (arable crops, breeding, livestock fattening); a technical revolution through the use of industrial inputs (pesticides, fertilisers) and genetically-improved seeds; and the intensification of livestock production (poultry, pigs and cattle) on the basis of industrially-processed feed (cereals, soybean).
>
> (Tubiana 1989: 26)

Following agricultural recovery in Western Europe, the continuing expansion of US exports centred on the rising demand for animal feed induced by the shift towards US consumption patterns. Within twenty years, beef consumption had doubled and that of poultry rose threefold. The grains-intensive livestock complex was successfully transposed, though with wheat replacing corn as the source of carbohydrate in compound feeds. This transfer had gained initial momentum under the Marshall Plan, which financed US soya bean shipments to the idle crushing plants of Northern Europe (Berlan et al. 1977). Aggressive export strategies by US soya bean interests subsequently consolidated this bridgehead, reducing the market share of ex-colonial oilseed producers, such as Senegal. These lost further ground after 1967, when their preferential trading arrangements with the former metropolitan powers were abolished.[1] The withdrawal of these vestigial privileges coincided with the introduction of the Common

Agricultural Policy (CAP) and agreement on a new division of labour in agricultural trade between Europe and the United States, which was negotiated during the Kennedy Round of 1964–7.

Under the compromise reached, EC grain and livestock sectors were protected by the system of variable import levies, which was instituted by the CAP to offset differences between world prices and EC prices. For oilseeds, on the other hand, the previous system of free trade was maintained, with preference given to soya beans and soya meal, and hence to the United States, the leading exporter. These products enter the European Community duty free, whereas other oilseeds and meal attract a 2 per cent duty. With this compromise, so advantageous to the American soya bean complex, a vital component for the modernization of European agriculture along US lines fell into place, with broadly similar consequences in terms of greater specialization, intensification and concentration of production.

THE DISARTICULATION OF THIRD WORLD AGRICULTURE

In the Third World, shortages of hard currencies, ever-mounting US agricultural surpluses and Cold War politics combined to establish a different trajectory for the diffusion of the US technology/policy model and consumption norms. The mechanism encountered represented a coincidence of interest, albeit temporary and one with strongly adverse long-run consequences for rural structures and food security in peripheral economies. In developing countries implementing accelerated industrialization programmes in the 1950s, cheap food supplies were needed to prevent rising real wages from reducing profits, and therefore discouraging accumulation. Cheap food was the *sine qua non* of cheap labour, the cornerstone of peripheral industrialization. Fordism was present only as a hybrid, severely distorted version of its Northern exemplar: fordist consumption goods, including meat-based diets, were produced not as wage goods for a mass market of the urban proletariat but for the higher income groups, while cheap labour policies characteristic of extensive accumulation were adopted, at times enforced by overt state repression. In these conditions of 'socially disarticulated accumulation' (de

Janvry 1981), low prices of staple foodstuffs were secured by reducing returns in family-labour agriculture, the traditional source of these wage goods, by implementing domestic price controls, subsidizing the technological modernization of large-scale producers and increasing food imports.

Plate 4.1 Indian women buying food from a government CONASUPO store, Mexico

These demands coincided with US efforts to rationalize management of its chronic farm problems by disposing of commodity surpluses overseas. This was done by greater recourse to subsidy payments to promote commercial exports, since domestic support prices typically exceeded world price levels, and by the gradual construction of a food aid policy (Friedmann 1982). The second element of this dual strategy, which grew out of the US experience of immediate war relief, post-war economic reconstruction and the Marshall Plan, was firmly established by the Agricultural Trade Development and Assistance Act of 1954, known as Public Law (PL) 480. According to Cochrane and Ryan, PL480 'marked a turning point in government export policies and programs, especially those dealing with concessional or non-commercial shipments' (Cochrane and Ryan 1981: 266). Not only were earlier surplus disposal and foreign food aid programmes consolidated, but 'This new legislation recognized both excess productive capacity in US agriculture and the dollar shortage in many food-short nations, impeding commercial imports from the United States' (1981: 266).

The new food aid policy was explicitly designed to circumvent this foreign exchange constraint: under Title I of PL480, 'concessionary sales' of surplus US commodities could be made to 'friendly' countries in return for inconvertible national currencies. In addition, these agreements were based on negotiated prices and highly favourable credit conditions, reflecting the long-term aim of developing new export markets. The foreign currency sales provision remained the key component of food aid policy, accounting for over 50 per cent of PL480 exports, until the Food for Peace Act of 1966, which gave priority to sales for dollars and convertible currencies. Other long-term dollar credit sales provisions reinforced this strategy of 'transforming aid recipients into commercial customers' (Cochrane and Ryan 1981: 274). These authors estimate that PL480 exports amounted to US$22.3 billion between 1954 and 1973, 20 per cent of total agricultural exports in this period, and contributed particularly to sales of dairy products, cotton and soya oils, wheat, and rice. In addition to this role in market creation, food aid strongly complemented US geo-political strategy in Latin America and the newly independent states of Africa and Asia, where the 'extension of wheat markets dovetailed with extension of the free world' (Friedmann 1982: 262).

In these conditions of easy availability and stable prices, food security and the vitality of domestic agriculture assumed secondary importance for many industrializing Third World countries. 'Urban bias' also was the conventional wisdom of development planning, which held that resource transfers from agriculture, mobilized via taxation, adverse internal terms of trade, and over-valued exchange rates, should be used to finance industrialization and support the strategic cheap food/cheap labour equation. The twin mechanisms of cheap food policy – domestic price controls and subsidized food imports – have severely weakened peasant agriculture, precipitating the present food crisis in the Third World. Discriminatory agricultural modernization policies are the second major cause of the decline of small-scale family-labour production and its correlate employment and self-provisioning structures. As a result, once agriculturally self-sufficient areas now experience rising levels of malnutrition, under-employment and poverty. The rural food crisis is an integral element of the Third World environmental crisis since it not only intensifies pressures on marginal ecosystems but also enhances the pivotal role of foreign exchange earnings from export agriculture in meeting domestic food requirements, further displacing staple food crops. This apparent paradox is at the root of the Third World food and environmental crises.

AGRICULTURAL MODERNIZATION: LATIN AMERICA

Cheap food policies and openly discriminatory agricultural modernization strategies, exemplified by those adopted in newly-industrializing Latin American countries, have resulted in sharply dichotomous patterns of income, productivity and innovation between sectors producing domestic staples and those specializing in exports, industrial raw materials and luxury food crops. Direct price controls, concessional and commercial imports and, especially before the mid-1970s, limited access to subsidized modernization programmes have depressed relative returns in staple food production. Conversely, priority export sectors have benefited from 'institutional rents', conferred notably by subsidized investment credits, as well as the introduction of more open trade regimes and export promotion policies since the mid-1960s. Intersectoral differences in profitability and risk therefore have widened over time, diverting land and resources away from staple wage goods.

The shortfall in domestic food supply is aggravated by low

rates of productivity growth and innovation as staple foodstuffs increasingly become the preserve of poor, under-capitalized family producers using traditional risk-averting practices in which only small areas are devoted to any one crop. These structural effects of institutional bias have severely constrained the capacity of staple food producers to adjust to subsidized imports and the associated shifts in urban consumption towards wheat products and animal protein. In Brazil, for example, where the daily per capita availability of calories and proteins, measured by an aggregate 'basket' of rice, beans, corn, manioc and potatoes, fell by 20 per cent between 1967 and 1979, this decline was partially met by subsidizing the domestic consumption of imported wheat. In this period, the share of wheat in protein availability rose from 14 to 25 per cent and from 12 to 21 per cent of calorie availability (Homem de Melo 1983). Similar trends emerge clearly at the regional level for the period 1964–84. Staple food production has increased relatively slowly, averaging 2 per cent annually, and its share in regional output has been eroded by the expansion of export crops and urban food products, such as coffee, citrus fruits, vegetables and eggs (IDB 1986). The share of oilseeds increased significantly, especially soya beans, whose output rose eighteenfold in this period, although this growth was heavily concentrated in Brazil and Argentina.

These changing land use patterns also have resulted in the rising concentration of land ownership due to the declining competitiveness of small family labour producers of staple foodstuffs. Crop substitution therefore has led to the increasing marginalization of traditional agriculture, accelerating processes of proletarianization and rural out-migration. Some small producers with a more favourable initial resource endowment have adapted to the demands of technological modernization through greater cash crop specialization but the majority has been reduced to semi-proletarian status. In Latin America, this disarticulation of traditional peasant agriculture is reflected in the large concentrations of impoverished rural households, which arguably is the outstanding feature of contemporary agrarian structures. According to Food and Agriculture Organization (FAO) estimates for the late 1970s and early 1980s, the incidence of absolute rural poverty reached 67–85 per cent, while the number of sub-family-sized farms has continued

to grow at an annual compound rate of 2.2 per cent in a group
of seventeen Latin American countries between 1950 and 1980
(de Janvry and Sadoulet 1988).

> It is likely that as much as two-thirds of the farm
> households across Latin America derive more than half
> their farm income from off-farm sources – principally
> wages from employment both in agriculture and in a wide
> variety of other activities.
>
> (de Janvry and Sadoulet 1988: 17)

Agricultural modernization strategies were a response of
rapidly urbanizing and industrializing countries to the
diagnosis, common in the 1950s and 1960s, that the low
productivity and inelastic supply allegedly characteristic of
'backward' or 'traditional' agriculture would inhibit growth
and development. Although this diagnosis was consistent with
reformist policies and the 'farmer road' of capitalized family
producers, such strategies, at least in Latin America, were
incompatible with the hegemonic class alliance or 'structural
pact' between industrial capital and the rural oligarchy of
landed *latifundist* interests. This alliance is at the root of cheap
food/cheap labour policies of industrialization.

Modernization policies also were prompted by fears of
diminishing returns to existing technologies, whether due to
growing population pressure and the fragmentation of
holdings or the 'closure' of extensive agricultural frontiers. The
transition to more capital- and energy-intensive methods was
facilitated by the available 'shelf' of Green Revolution technologies whose diffusion was actively promoted by the international scientific establishment and agro-industrial capitals
seeking to extend overseas markets. This 'technological fix' for
traditional rural structures also appealed strongly to US-trained
agricultural economists and advisers to bilateral and multilateral
aid agencies. These forces, despite occasional, though limited,
reformist initiatives, launched Latin American countries on the
road of 'Junker capitalist development' (de Janvry 1981), known
in Brazil as 'conservative modernization' since the inherited
latifundio-dominated agrarian structure was strengthened.

Leaving equity issues and social impacts aside, aggregate
indicators reveal that the transition to land-saving yield-
increasing technologies accelerated sharply in Latin America

after 1960. Studies of the sources of output and productivity growth emphasize the significant contribution of mechanization and fertilizers in the 1960s, reinforced in the 1970s by improved seeds, selective livetock breeding and better farm management (IDB 1986). Between 1969 and 1982 fertilizer consumption per hectare of land under crops grew by 6.9 per cent annually, rising from 22kg to 36 kg.

In the main, fertilizer consumption in the 1970s was associated with the use of improved seed varieties in grain production, especially in Mexico and Colombia, and the expansion in Brazil and the Southern Cone of such crops as sorghum and soybean, both based on technological packages that included the use of agricultural chemicals.

(IDB 1986: 103)

Excluding Brazil, which presents specific ecological problems, short-stem high-yielding varieties (HYVs) of rice are used in over 70 per cent of the area planted to this crop, notably in Mexico and Colombia. Dwarf HYVs of wheat account for more than 80 per cent of Latin American acreage, rising to over 95 per cent in Argentina, where wheat increasingly is grown in rotation with soya beans. Modern technological packages also have been developed for corn and sorghum hybrids (IDB 1986).

The structural effects of the modernization of rice production in Colombia, which followed the diffusion of improved varieties developed by the International Centre for Tropical Agriculture (CIAT), are instructive. In the thirty years after 1950, rice yields more than doubled from 1.8 tons to 4.0 tons per hectare, and total production had risen from 241,000 tons to over 2 million tons by 1982. Before the advent of Green Revolution HYVs, rice in Colombia was typically a peasant crop grown principally on holdings under 5 hectares. However, the number of farmers growing rice was halved in the 1960s alone, and almost 50 per cent of the area planted with rice was in holdings of 100 hectares or over. As in neighbouring Ecuador, the modernization of rice production in Colombia responded to urban needs for cheap food but exacted a high price in terms of rural proletarianization, greater land concentration and the heavy toll of intensive agri-chemical use on the physical environment (Redclift 1978).

Moreover, although rice became a staple – 'el pan del

143

pueblo' – in the diet of the poor, Colombia's food imports increased tenfold to over 1 million tons between 1953 and 1980. This example is symptomatic of the deterioration in the food security of the region as a whole. FAO projections suggest that Latin America's requirements for imported food will rise in the period from 1980 to 2000, when an additional 60 million tons in wheat equivalents will be needed. Yet, even before the onset of the debt crisis, over 50 million people had a daily calorie intake below the minimum recommended by the World Health Organization, and 20 million of these were considered to be seriously malnourished.

The debt crisis and macro-economic stabilization policies, whose adverse effects are felt disproportionately by low income groups, the unskilled and the poor, undoubtedly have aggravated what is already a chronic problem. Furthermore, in addition to slower economic growth, higher unemployment and cuts in social expenditures, macro-economic adjustment programmes supported by the IMF and the creditors' cartel frequently include measures to cap or reduce food subsidies (Cornia *et al.* 1987). Despite growing food insecurity in Latin America, external pressures to service the debt have maintained the institutional bias in favour of export crops at the expense of staple food production, adding to domestic food supply difficulties, as we shall see below.

The Brazilian case

Modernization policies in Latin America were implemented mainly through the expansion of rural credit at low, often negative, rates of interest, which strongly subsidized the cost of capital, extending 'institutional rents' to those with access to the system. Taking Brazil as our example, rural credit subsidies in 1978 were the equivalent of 3–4 per cent of gross domestic product, yet only 25–30 per cent of producers had access to these massive resource transfers (IBRD 1979). Moreover, the allocation of total rural credit, which rose fivefold in real terms between 1969 and 1978, was highly selective by crop, giving strong preference to export commodities (coffee, sugarcane, soya beans) and wheat, an import substitution crop, over staple foodstuffs. The large estate has been the privileged target and beneficiary of these policies, which have heavily underwritten the risks of innovation, both for current inputs and capital goods. With loans typically 'tied' to specific industrial inputs and dependent on the adoption of approved crop management practices, modernization became virtually synonymous with the 'industrialization' of the rural labour process, and hence with the expansion of the agro-industrial complex. Using these institutional mechanisms of transfer and subsidy, the state has mediated the integration of modernizing agricultural sectors with agro-industry. In Brazil, these linkages were reinforced by credit and fiscal incentives for investment in processing plants and marketing systems, the creation of 'export corridors' and major import substitution programmes in agricultural implements, machinery, and agri-chemicals.

This diversification process was inextricably associated with increasing transnational penetration of key sectors of the Brazilian agri-food system. A recent study (BNDES 1988) provides ample evidence of this internationalization, which has occurred to a lesser or greater degree throughout Latin America. In the mid-1980s, transnational firms and their affiliates held 80 per cent or more of the heavily concentrated markets for tractors (Massey Ferguson, Valmet, Ford) and harvesters (Ford, John Deere, Massey Ferguson).[2] The list of transnational corporations producing pesticides and other fine agri-chemicals is a roll-call of the firms which dominate the international scene, led in Brazil by Dow Chemical and Shell with 15 per cent each of the market. State and national private capitals are pre-eminent in fertilizer production thanks to the import substitution policies adopted in the 1970s. In the seed industries, Agroceres, now controlled by national capital but established originally by Rockefeller interests, has 45 per cent of the hybrid corn market, although transnational corporations are well represented by Ciba Geigy, Shell, Cargill, Continental Grain and Pioneer Hi-Bred International. Agroceres, in association with international groups, namely, Ross Breeders (poultry) and Pic-Pig Improvement Company, is the only national firm to acquire the genetic technology needed to produce the pure breeding lines used in large-scale poultry and pig production. Other local producers continue to import their breeding stock from oligopolistic international suppliers, such as Arbor Acres and Hubbard in the case of poultry (BNDES 1988: 32).

145

Transnational corporations are well entrenched in downstream sectors of the Brazilian food system, including Cargill and CPC International in the primary processing of oilseeds and corn, respectively. There is also a strong transnational presence in food manufacturing due to the activities of Nestlé, BSN, Ferruzzi, Bunge-Born, RJR Nabisco and others. Finally, reflecting international experience, national groups are predominant in food distribution, although catering has attracted transnational hotel firms and fast-food chains, such as McDonald's, to Brazil's metropolitan areas.

Modernization has brought dramatic changes in rural labour markets following the internal reorganization of the rural production process to incorporate labour-saving innovations. Even where irrigation permits multiple cropping, mechanization and the use of herbicides have reduced the demand for permanent, resident labour and accentuated its seasonality (Maffei *et al.* 1986). In Brazil, these effects have been magnified by the substitution of mechanized, short-cycle crops, such as soya beans, and permanent pasture for labour-intensive crops, such as coffee. Estimates for 1975–7, for example, indicate that coffee required 83 man-days per hectare, whereas soya beans needed only 3 man-days (Goodman and Redclift 1981). In south-eastern Brazil, where large estates predominate, subsidized modernization has accelerated the proletarianization and 'casualization' of the rural labour force. Permanent, resident workers and 'internal peasantries' of sharecroppers and other tenants have been expelled and replaced by casual wage workers and specialized, semi-skilled labour, with landowners exercising more centralized management and control of production. The expansion of a landless, casual rural proletariat led to the emergence in the 1970s of *favela* settlements on the periphery of small rural towns, a phenomenon hitherto only associated with large metropolitan centres. These settlements are a direct reflection of the rural food crisis associated with new patterns of accumulation within agriculture and the wider agri-food system.

Agricultural modernization strategies have similarly transformed the conditions of reproduction of independent 'external' peasantries. In some cases, this has occurred through the direct integration of small family producers into modernization programmes, leading to higher capitalization and

146

THE THIRD WORLD FOOD CRISIS

greater cash-crop specialization, which increases their vulnerability to market fluctuations. In Brazil, the transformation of diversified, self-provisioning family agriculture is exemplified by the wheat–soya bean sector. This was the target of wheat import substitution programmes in the late 1950s and the subsequent expansion of institutional credit for soya beans to support Brazil's spectacular entry into the world market in the late 1960s. Many of these capitalized small producers have been trapped in a 'reproduction squeeze' due to their unsuccessful efforts to finance the continuing capitalization process, resulting in chronic indebtedness, forced land sales and eventual proletarianization.

Direct integration with agro-industrial capitals is also illustrated by contract production and the consequent loss of effective control over operating decisions. In other cases, the modernization of large estates may reduce opportunities for seasonal wage employment and the renting in of land, which are vital to the reproduction of small producers. This symbiosis between small properties and large estates, known as the *latifundio–minifundio complex* in Latin America and articulated through local labour markets and tenure relations, has suffered a crippling blow from modernization programmes, accentuating rural poverty and reducing the capacity of the rural population to feed itself and its urban counterpart. In northeast Brazil, for example, the widespread conversion of large properties to permanent pasture has intensified social differentiation and outright proletarianization, although large concentrations of impoverished small producers remain.

INTEGRATED RURAL DEVELOPMENT PROGRAMMES

A new chapter in the transformation of small-scale family agriculture opened in the 1970s, when US foreign agricultural policy under Nixon gave greater emphasis to commercial exports, reducing food aid and concessionary sales. This change hardened with the sharp decline in world grain stocks which followed the huge Soviet grain purchases in 1972 and the poor US harvest of 1974, reviving neo-Malthusian fears of global famine as wheat prices jumped from $2 to $4.90 a bushel between 1972 and 1974. These events turned attention towards domestic food supplies and the limitations of modernization policies. These had achieved significant rates of expansion for export crops, industrial raw materials and high-value products but at the cost of displacing staple food crops.[3] As we see below, an increasing asymmetry emerged between domestic supply structures and the food needs of the mass of the population. Major export sectors (soya beans, winter vegetables, beef) are only partially integrated with mass consumption patterns, while food and feed grains increasingly are consumed indirectly in the form of meat, which is beyond the limited means of the rural and urban poor. In Mexico, for example, even the social profile of wheat consumption is highly stratified as it is the basic food of the industrial workforce and higher income strata, yet plays little part in the diet of the poor, who are predominantly rural dwellers (Sanderson 1986). We return to these asymmetries and dualities when discussing the food crisis in greater detail.

Integrated rural development programmes (IRDPs) were the major institutional focus of this renewed concern for domestic food supplies, and hence small producers. This initiative, launched in 1973 when the World Bank announced a worldwide funding programme for IRDPs, tacitly accepted the limitations of modernization programmes and critiques of the distributive bias of Green Revolution innovations. IRDPs were seen as a second stage in the diffusion of Green Revolution technologies, which had to be more tightly managed and 'targeted' if small producers were to gain access to institutional credit and research and extension systems. Even so, IRDPs have been criticized for excluding the mass of small producers and concentrating on those whose initial resource endowment made technical innovation and increased commercial production attractive. Experience with IRDPs also has revealed the wide gulf that separates officially recommended technological 'packages' and small farmer practices, including their approach to risk, and the diverse conditions found in resource-poor environments, where rural poverty is heavily concentrated. Since the 1960s, the international agricultural research community gradually has sought to accommodate its critics by advocating a wider, farming systems approach, although for the most part the 'top-down' orientation remains.[4] The real challenge is to develop more productive but sustainable agricultural systems for less well-endowed environments. This goal is paid lip service but, in the meantime, the traditional commodity-based focus continues to dominate both the agricultural research agenda and policy efforts to mobilize commercial surpluses in peasant agriculture.

The neglect of low-resource farmers' needs is due to many factors, ranging from the balance of political forces in the formulation of agricultural development policy to the reductionism of mainstream agricultural research and extension systems. These typically are dominated by the 'transfer-of-technology' model in which 'scientists largely determine research priorities, develop technologies in controlled conditions, and then hand them over to agricultural extension to transfer to farmers' (Chambers and Jiggins 1987: 35). The effectiveness of this model depends critically on farmers' access to relatively uniform, resource-rich environments, sources of modern inputs, and rural infrastructure, creating positive feedback effects between research, crop productivity and farm incomes. In terms of the 'transfer-of-technology' model, where these conditions are met, parallels can then be drawn between grain production in the American Mid-West, the Beauce plain of northern France and the Indian Punjab, for example. However, since the millions of poor farmers do not, by definition, enjoy access to these favourable conditions, the 'packages' developed by agricultural research stations are largely irrelevant to their needs and priorities. In contrast with the high productive potential of such 'core' areas as the alluvial plains of South and South-east Asia, 'the hinterlands in which many RPFs [resource poor farmers] are found are highly diverse geomorphologically, ecologically and culturally, demanding highly differentiated and locale-specific research' (Chambers and Jiggins 1987: 37). Such research, moreover, must also be situated within a wider political economy which directly addresses issues of equity, class conflict and sustainability. These questions remain outside the conceptual and methodological framework of farming systems research, which is concerned primarily to integrate 'forgotten' small producers into the market (Brouwer and Jansen 1988).

TECHNOLOGY DIFFUSION AND
THE GREEN REVOLUTION

We have seen that corn hybridization in the United States triggered the convergence of the major historical tendencies of industrial appropriation to produce high-yielding, fertilizer-responsive and machine-harvestable varieties. The technical capacity to modify the basic biological characteristics of major commercial crops, allied to the increasing privatization of plant breeding research, transformed hybrid varieties into truly *industrial* seeds, opening up a new terrain of accumulation. This nexus between agri-genetic innovation and profit, reinforced by calculations of political advantage, was the driving force behind the Green Revolution, which internationalized the crop-breeding methods first applied in the Mid-West Corn Belt.[5]

As has been argued elsewhere, the extension of these techniques to tropical and sub-tropical environments in the Third World 'marks the further homogenization of the agricultural production process around a common set of agronomic practices and generic industrial inputs' (Goodman, Sorj and Wilkinson 1987: 45).

In order to realize the full genetic potential incorporated in these 'delicate custom-built plants' (Pearse 1980), industrial inputs are necessary to reduce environmental stress and establish favourable growing conditions. High productivity thus depends on capital-intensive soil management practices (fertilizers, agri-chemicals, irrigation) to create controlled fertile environments for these carefully selected varieties. Not surprisingly, international agribusiness was well to the fore in the diffusion of Green Revolution technologies, profiting from state-subsidized markets for modern inputs to extend its influence over local agri-food systems. The emergence in larger Third World countries of 'agro-industrial complexes', dominated by transnationals and their subsidiaries, was accompanied by parallel trends in downstream food manufacturing sectors as patterns of food consumption increasingly became internationalized. As we observe below, the modernization of Third World agriculture has replicated developed country experience in other ways as the new 'miracle' seeds have encouraged the growth of monocultural systems, displacing indigenous varieties and increasing genetic vulnerability.

The first mechanisms of technology transfer combined private philanthropy and US government funding. The Mexican Agricultural Programme for wheat and corn improvement was established in 1943 with Rockefeller Foundation support, apparently with the blessing of Henry A. Wallace, then US Vice-President and formerly Roosevelt's Secretary of Agriculture and founder of the successful corn seed firm, Pioneer Hi-Bred (Kloppenberg 1988a).

Plate 4.2 Farming systems researcher conducting trials on small farm, Guatemala

In the later 1940s, this initiative was extended to other Central American and Latin American countries by the US Department of Agriculture (USDA) and American land-grant universities, a pattern subsequently followed by various funding agencies elsewhere in the world for a growing range of crops. This institutional development was consolidated in the 1960s when the Rockefeller and Ford foundations supported the creation of four specialized international agricultural research centres (IARCs) devoted to selected crops, beginning in 1959 with the International Maize and Wheat Improvement Centre (CIMMYT) in Mexico. The IARC network, which provided the main channel for the transfer of plant breeding techniques and the dissemination of HYVs, now includes thirteen centres and is funded by an international consortium of Organization for Economic Co-operation and Development (OECD) governments, multilateral institutions and private foundations. The task of adapting improved germplasm to local environments usually is undertaken by national agricultural research and extension agencies, often established initially with World Bank support, and modelled on the US system. In 1971, the international donors funding the IARCs vested responsibility for the co-ordination and development of the network in the Consultative Group on International Agricultural Research (CGIAR). The role of the CGIAR in the collection, evaluation and conservation of plant germplasm has become increasingly controversial, as we shall see in Chapter 6. There are also widespread doubts that the CGIAR will provide an effective conduit for the transfer of plant biotechnologies to the Third World.

THE NEW INTERNATIONAL DIVISION OF LABOUR

We have stressed that cheap food policies, based on food aid, consumer subsidies and strongly discriminatory agricultural modernization policies, displaced traditional crops, such as corn, millet, manioc and sorghum, as wage foods in developing countries. Moreover, a rising share of these crops is processed as animal feed to meet expanding middle-class demand for livestock produce. Following the US decision after 1972 to reduce food aid and concessionary exports, developing countries have turned increasingly to agricultural exports in order to finance food imports. These efforts often have focused on industrial crops or luxury high-value products, widening the gap between domestic production and local requirements. These tendencies, which have been particularly pronounced in middle-income industrializing countries and oil exporters, have brought significant changes in the volume and direction of agricultural trade. The outstanding feature of the new international food order is the dependence of many Third World countries on food imports to meet their basic needs. Fundamentally, however, the food crisis in developing countries, with few exceptions, arises from the regressive distribution of income, which finds expression in patterns of land use dictated by export markets and the demand of higher income groups.

Table 4.1 World grain trade 1934–80 (million tonnes)[1]

	1934–8	1948–52	1960	1970	1979–80
North America	5	3	39	56	127
Western Europe	−24	−22	−25	−30	−13
Eastern Europe and USSR	5	n.a.	0	0	−40
Australasia and New Zealand	3	3	6	12	15
Japan	−2	−3	−4	−14	−23
Latin America	9	1	0	4	−9
Africa	1	0	−2	−5	−13
Asia[2]	0	−3	−13	−23	−39

Source: Grigg (1986) based on Hopkins and Puchala (1978) and FAO data.
Notes: 1 In million tonnes. Net exports and net imports are shown by positive and negative values, respectively.
2 Excluding Japan

In the twilight of the pre-war international food order, Latin America, Africa and Asia were largely self-sufficient in food grains. The agricultural trading system before 1945 was structured around Western European demand for cereals and specialized export production in the independent settler states of the Americas and Australasia (Table 4.1), together with the older metropolitan-colonial commerce in tropical products. In the 1950s, Third World self-sufficiency in cereals succumbed to the combination of policy interventions in domestic and international markets noted earlier, which significantly distorted the effects of population growth and rapid urbanization on demand for cereals. Indeed, in the two following decades, the middle-income group of developing countries fed one-third of domestic grain production and half their cereal imports to livestock (Grigg 1986). The roots of the new import dependence were nurtured by the period of exceptionally low world prices arising from US farm policies and America's domination of the world grain trade, which it asserted by dumping domestic production surpluses through PL480 sales and other subsidized export programmes (Friedmann 1982).

> Gross [wheat] imports of the countries in Latin America, Asia and Africa, excluding Argentina and Japan, had risen to more than a third of the world total by 1959. These accounted for almost half of world imports in 1971, on the eve of the first large Soviet purchases.
>
> (Friedmann 1982: 265)

The dramatic rise in world wheat prices which followed these events in the mid-1970s failed to change the fundamentals of the new international division of labour established under American hegemony. Third World consumers were insulated from rising international prices by widespread use of subsidies on wheat and wheat products. In fact, despite rising prices and greater market instability, world trade in food products grew at 5.3 per cent a year at constant 1980 prices during the period from 1971 to 1984, outstripping manufactured exports. The lion's share of the expansion in world food exports was taken by the industrial market economies, which accounted for 46 per cent in 1961–3 and 63 per cent in 1982–4 (IBRD 1986). Demand was sustained principally by the middle-income developing countries, which absorbed 80 per cent of the growth in Third World food imports between 1962 and 1984. These imports continued to grow rapidly after 1975, financed by increasing foreign indebtedness and, particularly in North Africa and the Near East, by 'oil rents' from OPEC price increases. Despite the constrained financial conditions of the 1980s arising from the debt crisis and the subsequent collapse of world oil prices, efforts to alleviate import dependence by cutting food subsidies were undermined by political pressures, including 'bread riots' in several North African countries.[6]
These shifts in the structure of consumption and the

direction of trade are exemplified by the rise of the developing countries' share of world wheat consumption from 39 per cent in 1963–4 to 49 per cent in 1979–81 and 53 per cent in 1985 (IBRD 1986; International Wheat Council 1988). Wheat consumption in developing countries grew at an annual average rate of 5.5 per cent in the years from 1975 to 1985, compared to only 1.6 per cent in the industrialized countries (International Wheat Council 1988). By 1985, developing countries' gross wheat imports averaged 57 million metric tons, representing 63 per cent of world trade in wheat, with the industrialized countries occupying an overwhelmingly dominant position as exporters. Indeed, Teubal (1987) cites USDA estimates that roughly 20 per cent of US farm acreage in 1985 was devoted to crops for export markets in the Third World.

These developments are part of the severe food crisis which now grips Africa where, with few notable exceptions, per capita food production has fallen since 1970. Moreover, the rapid growth of imports, whose share in total calorie consumption doubled from 6 per cent in 1967–71 to 12 per cent in 1979–81, failed to prevent per capita food consumption in sub-Saharan Africa from declining during the period between 1960 and 1984 (Platteau 1988). Since 75 per cent of the population in this region is engaged in smallholder food production, malnutrition and chronic poverty are direct consequences of the virtual stagnation of output. Despite the declining trend of per capita cereal consumption, Africa also has experienced what Yotopoulos (1985) terms 'food–feed competition' and the use of major food crops for animal feed has increased significantly. Platteau (1988) joins a wide consensus in attributing these substitution effects, including the rise of wheat and rice at the expense of traditional staples (millet, sorghum), particularly in urban diets, to the artificially low levels of consumer food prices 'due to overvaluation of exchange rates, subsidies, international food aid . . . (Platteau 1988: 7).[7]

Unlike Latin America and Asia, where the poor more often are net food buyers, such policies in Africa discriminate directly against poor rural households, who typically are net food *sellers*. These policies have 'By contrast, enriched urban middle classes [who] can all the more easily shift to animal protein diets as the cost of feed is artificially low' (Platteau

1988: 7). Recent research by the International Food Policy Research Institute (IFPRI 1987) also draws attention to the impact on dietary habits of urbanization and new employment opportunities. Thus the sharp rise in urban rice consumption in West Africa since the early 1960s reflects new work patterns of urban households: 'As women enter the work force, and men work away from home, there is strong demand for staples that can be prepared quickly at low cost and that are available in roadside restaurants. Rice meets both needs' (IFPRI 1987: 1–4).

A recent International Wheat Council report for the period from 1975 to 1985 adds further detail to these trends and changing consumption patterns in the major regions of the Third World. Shifts away from coarse grains as staple foods were particularly evident in India and sub-Saharan Africa, where per capita consumption actually declined. The feed use of grains increased in every developing region, although this growth is from a low base, as in the case of sub-Saharan Africa. This region also exemplifies the strong influence of urbaniza- tion and new lifestyles on dietary patterns, represented by the shift towards 'processed products like milled rice or flour, or ready-to-eat foods, especially wheat bread' (International Wheat Council 1988: 6). Urban populations in the low-income countries of sub-Saharan Africa are rising at an annual rate of 5–6 per cent, and total wheat consumption has doubled in each decade from 1 million tons in the mid-1950s to 7 million tons in 1985. 'Nearly the whole of this quantity was consumed in the cities, and 80 per cent of it was imported' (1988: 6). Continuing rapid urbanization suggests that these substitu- tion effects between staple and imported grains will intensify over the next few decades. According to middle-range projections by the United Nations, the urban population of sub-Saharan Africa will grow from 127 million in 1985 to 763 million by 2025, when it will represent 56 per cent of the region's total population (Goliber 1989).

The impact of urbanization on diets is seen at a more localized level in the town of Kingston, Jamaica. Here, rural migrants within twelve months of arrival significantly increase their consumption of rice, wheat bread, chicken and eggs at the expense of such 'country' staples as yams, bananas and sweet potatoes, which subsequently are abandoned by the new city-dwellers (Paquette 1984). This 'Westernization' of urban

156

food systems is particularly evident in the great port cities of
Asia, Africa and Latin America (Charvet 1987). In Senegal, for
example, an inhabitant of Dakar consumes 10 kg of corn, 77
kg of rice and 33 kg of wheat annually, whereas his peasant
cousin on average consumes 158 kg of corn, 20 kg of rice and
2 kg of wheat (Benoist 1983). As Charvet (1987) wryly observes,
wheat consumption has grown rapidly among the inhabitants
of eco-systems where it is scarcely possible to grow this crop.

This discussion of major post-war shifts in the volume and
direction of trade in food and feed grains focuses on North–
South flows and the adjustment of Third World countries to
growing import dependence, typically by promoting export
crops rather than strengthening staple production. These
trends capture a significant dimension of the internationalization of Third World agriculture but this analysis is centred too
exclusively on the nation-state and its policies. This approach
must be extended to incorporate the central, catalytic role of
the transnational agri-food corporation in disseminating
standardized technologies and labour processes and integrating
commodity markets on a global basis. Internationalized production technologies and the world-wide 'sourcing' of inputs
common in manufacturing have their counterpart in the agri-
food industry, with direct repercussions on land use, food
security, labour processes and the social relations of
production.

THE TRANSNATIONALIZATION OF THIRD
WORLD FOOD SYSTEMS:
LATIN AMERICAN CASES

Transnational corporations (TNCs) are the vector linking
regional economies and crop sectors in the Third World with
global systems of production and consumption.[8] New
products and new modes of integration have emerged which
complement and often eclipse traditional agricultural exports.
One illustrative example is the organization of beef production via forward contracts between calf producers in northern
Mexico and feedlots in Texas and California (Sanderson 1986).
In this case, 'there is an authentic international division
of labour, in that the final product comes from sequential
production in ''parts'', ''components'' or ''inputs'' in two

countries with very different productive specialization' (Rama 1985: 73). Other recent examples of transnational agro-industrial integration include exports of Mexican winter vegetables to the United States, Chilean fruit to Europe, Central American meat packing for US hamburgers, oilseeds and feed grains for European livestock from Argentina, Brazil and Thailand, and Brazilian exports of chicken and frozen orange juice concentrate.

Without exploring these cases in detail, the mechanisms which distinguish the new, agro-industrial modes of integration include the competitive standardization of production processes arising from global procurement policies, which may specify the cultivation practices to be used and stipulate international quality standards and other marketing criteria. Increasingly uniform patterns of consumption and tastes accompany and reinforce this tendency. Sanderson (1986) observes, for example, that all exports of fresh citrus from Mexico pass through US Department of Agriculture (USDA) certified packing stations, while beef exports from Latin America must meet USDA standards of hygiene and slaughter. Standardized production technologies also result from the market power exerted by agro-industrial processors and international buyers, whether through 'closed' vertically integrated systems, out-grower arrangements, or the many forms of contract production. Thus, 'In the case of Mexican winter vegetables . . . producer technology, financing and sales all transcend mere episodic market interaction and involve contract relationships, transnational brokerage and consignment sales to prearranged markets. The same is true of strawberries' (Sanderson 1986: 52). Moreover, despite the relatively long experience of the Mexican frozen vegetable and strawberry industries, innovations in these sectors depend on the R&D of Californian growers and transnational seed companies, which emphasizes their vulnerability in the United States, their principal export market (de Janvry et al. 1987).

In recent years, TNCs have penetrated all segments of Third World food systems but their direct presence is most significant in commodity marketing and brokerage, contract production, primary processing, food manufacturing and catering, including fast-food outlets. This concentration reflects a strategic shift towards more dynamic, higher value-added downstream sectors, as well as both forced and voluntary divestment of land holdings in plantations industries. Although the influence of TNCs extends to all Third World regions, their presence is especially marked in Latin America, where US agribusiness has been active since the inter-war period. In addition, information is more readily available on US companies and their affiliates in Latin America. These activities are put in wider perspective by emphasizing that perhaps two-thirds of international investment in the food system occurs within OECD economies. With notable exceptions, such as the Argentinian firm, Bunge Born, the hundred largest agro-industrial companies are located primarily in the United States (thirty-eight firms) and the United Kingdom (twenty-five firms) and control an estimated one-quarter of world food production (IAMM 1987). Against this background, we turn now to examine TNC activities in several different sectors of Latin American food systems.

The milk processing industry in Latin America has largely been controlled by transnationals since the 1930s when Nestlé and Carnation began to set up plants. This control frequently has been accompanied by monopsonistic power over supplies of raw milk. In Mexico, for example, Nestlé produces all the condensed and powdered milk and Carnation, which was taken over by Nestlé in 1985, controls supplies of evaporated milk (Teubal 1987). Urban demand for more sophisticated milk products, including yoghurt and European-style cheeses, has stimulated the entry of new TNCs, such as Gervais-Danone, into the industry since 1970 (Oman and Rama 1988). The yoghurt market in Mexico is virtually in the hands of only three firms: Gervais-Danone, Borden, and the Nestlé brand 'Chambourcy'.

Rising dependence on food imports and increasing transnational interests in domestic agro-industry were pronounced characteristics of the development of the Peruvian food system in the 1960s and 1970s. Agro-industrial growth was concentrated in the dairy and milling industries, edible oils and fats, confectionary, brewing and intensive poultry production (Caballero 1984). According to estimates for 1973 by Lajo (1978)

Bunge, Nicolini, and Cogorno groups controlled 88 per cent of the production of wheat flour used for bread and pasta; the same Bunge, Nicolini and [Ralston] Purina produced 80 per cent of the animal feed, the basis of the poultry and egg industry; again, Bunge, Pacocha [Unilever], and Romero controlled 64 per cent of oils, edible fats, and margerines, and Carnation and Nestlé controlled 100 per cent of evaporated milk.

(Quoted by Caballero 1984: 10–11)

This domination is accompanied, as in other Third World countries, by the transnationalization of tastes and consumption habits.

French-style bread is produced from hard wheat, 99 per cent of which is imported. Its enormous popularity (like that of pasta, which is also produced with imported wheat) has taxed heavily the import bill and has also impeded the participation of the Andean peasants in supplying the growing national demand for bread.

(Caballero 1984: 13)

TNC participation in Latin American livestock sectors has changed radically since the 1960s as the long-established 'majors', such as Swift, Armour and Wilson, have moved out of cattle ranching and fresh-meat packing towards processed meat products with higher value added and greater scope for brand differentiation (UNCTC 1981). TNCs increasingly have concentrated on processed pork and poultry products, often linked via interests in feed grain milling into vertically integrated production systems. In the beef industry, the move towards processed products was paralleled by a shift in the geographical locus of TNC activity away from the Southern Cone and into Brazil and Central America . In these countries, recent TNC investment has taken the form of minority participation in joint ventures to produce lower quality meats for processing abroad (UNCTC 1981; Oman and Rama 1988). The new ranching areas of Central America often are integrated through contract arrangements with fast-food chains in the United States, establishing the widely known 'hamburger connection' (Myers 1981).

This connection was very actively promoted in the 1970s by multilateral development agencies. Indeed, despite the furore aroused by recognition of the social iniquity and damaging environmental consequences of these lending policies, the World Bank continued to fund cattle ranching projects in Panama and Honduras in the early 1980s. Although beef production is the leading culprit, but also because of other export crops, such as sugarcane, bananas, pineapples, coffee and oil palm,

> almost two-thirds of Central America's lowland and lower montane rainforest have been cleared or severely degraded since 1950. At the current rates of destruction, most of the remaining forest will be eradicated in the next 20 years, leaving only impoverished remnants in national parks and reserves.
>
> (Nations and Komer 1987: 161)

The export-led momentum of this wantonly destructive process is brought home by evidence that, although two-thirds of Central America's arable land is now devoted to cattle ranching, beef consumption per capita has declined (Nations and Komer 1987). Narrow domestic markets mean that 25 per cent of Central American beef production is exported, with considerably higher ratios in Costa Rica and Guatemala. These exports go mainly to the United States, where the shift to grain-fed beef production created shortages of the leaner, grass-fed beef preferred for use in hamburgers and processed meat products. The links between the highly structured international market, trends in 'Northern' consumption patterns, and large-scale environmental devastation are captured vividly in the following passage:

> Central America's beef exports enter the United States in 60-pound boxes of frozen, de-boned cuts packed in USDA-inspected abattoirs in the Central American countries. Transported by refrigerated container ships to port cities in Florida and California, the meat is purchased by customhouse brokers and meat packers, then sold to fast food chains and convenience food processors throughout the country. Most of it is transformed into luncheon meats, hamburgers, frankfurters, chili, soups, beefstew, hash, sausages, TV dinners, frozen pot pies, baby foods, and pet foods.
>
> (Nations and Komer 1987: 165)

The presence of foot-and-mouth disease in Brazil has attenuated the US 'hamburger connection' with the felling of the Amazon rain forest, although processed beef products are exported to the United States. However, this disease has not prevented the rapid growth of Brazilian beef exports, principally to Western Europe. In the case of Amazonian deforestation, the impact of the internationalization of the agricultural resource base is more indirect, though nevertheless equally pervasive. It is seen in the modernization and concentration of agricultural production in the Centre-South, the concurrent rise of export crops, such as soya beans and citrus products, and the expulsion of masses of land-hungry rural workers to the rain forests of Brazil's vast northern periphery (Hall 1989; Goodman and Hall 1990).

The environmental crisis in Amazonia, as elsewhere in Latin America and the Third World more generally, is manifest in unsustainable agricultural production systems involving habitat destruction and the degradation of soils. However, its *social* content is the struggle for access to the resources essential for day-to-day survival. When viewed as livelihood and social reproduction, the environment in the Third World is a highly contested terrain and, although internationalization is a large part, it is not the whole story.

FOOD SECURITY

The internationalization of Third World agri-food systems, manifest in food aid, agricultural modernization and agro-industrial integration, has had an extremely corrosive impact on food security in many developing countries. Staple food sectors have been undermined simultaneously on several fronts, including shifts in domestic consumption induced by cheap-food policy, discriminatory modernization programmes to promote import substitution, notably of feed grains and export crops which bear little relation to local needs. Mexico, once extolled as an exemplary model of Third World agricultural development, became a food-deficient nation in the mid-1970s as the fulcrum of its agri-food system shifted decisively towards the livestock complex and export crops. Output of basic food grains (wheat, corn, rice) and beans declined in per capita terms between the mid-1960s and late 1970s as acreage previously devoted to these crops in rainfed regions was switched to oilseeds, principally soya beans, and forage crops (sorghum, barley, alfalfa). Corn, the staple *par excellence* of the poor, suffered particularly from this crop displacement in the rainfed zones, whereas wheat production moved to the irrigated areas of northern Mexico. In addition, an increasing share of corn production is diverted to the compound feed industry and consumed not by the rural poor but indirectly as meat by upper-income groups. Despite the spectacular growth of domestic production for the feedgrain–livestock complex since 1960, sorghum and soya beans each represented roughly a fifth of Mexico's agricultural imports in 1978–80 (Rama 1985: 80). This link via crop displacement between the rising ascendancy of the livestock complex and import dependence in basic foods is established to a greater or lesser degree in other countries of Latin America and in Africa and Asia.

Import dependence mirrors the growing divergence between domestic production and the consumption needs of the poor. The agricultural resource base, in other words, is increasingly geared to the demand of higher-income urban consumers and overseas markets. Pursuing the example of Mexico, the food crisis, defined in terms of nutritional deficiencies, is concentrated among the 24.5 million rural inhabitants, 39 per cent of whom suffered 'grave caloric deficit' in 1980 (Sanderson 1986). Productive resources, including peasant holdings previously used for self-provisioning, have been diverted to luxury crops, such as frozen strawberries, and feed grains, particularly for poultry, which are excluded from the diet of the majority. 'In Mexico, virtually no one in the lowest income strata eats poultry, eggs or any other animal protein on a regular basis' (Sanderson 1986: 56–7). Sanderson also observes that frozen orange juice concentrate produced in Mexico and Brazil, the world's second largest producer, is beyond the reach of the poor because they lack both income and refrigeration.

In important respects, Mexico and Brazil possess what can be called 'proto-fordist' agri-food systems, in so far as they exhibit an increasing, though still incipient, tendency for cereals to be consumed indirectly in the form of animal protein. In France and the United States, for example, indirect consumption accounts for 75 per cent and 90 per cent of total cereal consumption, respectively (Bertrand 1985). Brazil has started down this path, as shown by the rise in poultry consumption, the symbol of a 'fordist' diet, from 2.4 kg per capita in 1964 to 10.2 kg in 1982 (Bertrand 1985). Nevertheless, cheap wage strategies and the highly regressive distribution of income continue to weaken internal articulation between the feedgrain–livestock complex and mass consumption. This distortion is reflected in Brazil's position as a leading world exporter of soya beans since the early 1970s and, more recently, of poultry. Although soya bean oil, thanks in part to price controls, has become the main cooking oil in Brazil, about 75 per cent of soya meal is exported, indicating closer integration with 'fordist' consumption in Western Europe than domestically. The expansion of the area devoted to soya beans from 200 thousand hectares in 1960 to 10 million in 1984–5 reveals the magnitude of the internationalization of the productive base.

CONCLUSION

The structural maladjustment between an internationalized, export-orientated resource base and domestic food requirements which defines the food crisis in the Third World affords a sobering prospect. Malnutrition and rural poverty are immediate manifestations of this asymmetry. However, as Sanderson warns in the case of Mexico, the food crisis 'must also be considered as an early phase of future difficulties in feeding the growing rural population with an agricultural base less attentive to their needs and a trade base more constrained by external imbalances' (Sanderson 1986: 8).

This discussion clearly locates the origins of the Third World food crisis in the new mode of integration of agriculture in intensive ('fordist') regimes of capital accumulation, epitomized by New Deal farm programmes. The crop surpluses these generated supplied the food aid shipments and subsidized grain exports used to exert US domination over the post-war international food order. This order gradually disintegrated after the mid-1960s as US farm policy changed direction to give greater priority to commercial exports. Its final demise occurred abruptly under the weight of the US–Soviet grain deal of 1972 and poor harvests, which dramatically reduced world grain stocks. Wheat prices rose threefold in 1971–4, raising the spectre of global food scarcity and widespread famine in poorer countries.

After two decades of low stable prices, these developments exposed the 'hidden' food crisis in the Third World and the disarray of national food policies. Within a single generation, self-sufficiency in food grains had been squandered, at an incalculable cost to the nutrition and well-being of future generations. Moreover, import dependence effectively transfers control over natural resources to international markets and agri-food complexes dominated by a handful of powerful corporations. Agricultural sectors are mortgaged to earn foreign exchange and service external debt accumulated, in part, to pay for imports of basic foods.[9] The environmental consequences of this insertion of Third World countries into the new international division of labour is a leading theme of Chapter 6.

NOTES

1 In France, for example, the share of soya meal in animal feed has risen from 7 per cent in 1952 to 42 per cent in 1965 and 79 per cent in 1980, replacing other oilseeds, particularly peanuts, whose share fell from 34 per cent to 10 per cent in the same period.

2 National capital (Grupo Iochpe) acquired the controlling interest in Massey Ferguson's Brazilian affiliate in 1986, which gave it the leading market share of 35–40 per cent in each of these sectors.

3 This stricture is more applicable to Africa and Latin America than Asia, where Green Revolution technologies have contributed to the 27 per cent rise in per capita food production since the mid-1960s. In general, however, poor farmers with access to less well-endowed environments have fared badly. As Conway and Barbier observe, 'the Green Revolution has had significant problems of equity and in the stability and sustainability of production' (Conway and Barbier 1988: 652).

4 Conway and Barbier note that farming systems research has taken several different directions in the international and national agricultural research systems, which has resulted in 'a mixed record of success and failure in terms of providing significant benefits to farmers on resource-poor lands' (Conway and Barbier 1988: 663).

5 An earlier transfer of hybrid breeding techniques to tropical agriculture occurred in the Far East when Japan applied methods used in temperate rice cultivation in the nineteenth century to develop fertilizer-responsive varieties in Taiwan after 1900. This programme later provided germplasm used by the International Rice Research Institute (IRRI) in breeding IR-8, the first of the 'miracle' HYVs of rice. (Hayami and Ruttan 1971; Johnston and Kilby 1975; Pearse 1980).

6 For discussion of the situation in Tunisia and Morocco, see Seddon (1986).

7 In fairness, Platteau (1988) gives greater weight to historical legacies and underlying structural constraints in explaining the African food crisis and is critical of the dominant view, which focuses on 'faulty' pricing policies.

8 Clairmonte and Cavanagh (1982) estimate that 60–90 per cent of world trade in eight leading primary food products – wheat, sugar, coffee, corn, rice, cocoa, tea and bananas – is controlled by fifteen TNCs in each sector, with only three to six traders accounting for 'the bulk of the market' in most cases.

9 The international debt crisis has intensified efforts to expand agricultural exports. In the case of Brazil, for example, agricultural products accounted for roughly 40 per cent of total exports in 1984–6, and net agricultural exports represented around 70 per cent of the trade surplus in this period. Significantly, the net surplus on agricultural trade of US$9 billion in 1984 and $8 billion in 1985 amounted to 60 per cent of Brazil's *total* debt service payments in these years.

5

ENGINEERING LIFE: AGRI-BIOTECHNOLOGIES AND THE FOOD SYSTEM

In Chapter 3, we argued that the structural specificities of the modern food system are legacies of industrial capitals' inability to subordinate agricultural production and food supply to direct industrial transformation. The concepts of appropriation and substitution were introduced to analyse the strategies industrial capitals have followed to reduce the constraints on accumulation presented by the biological cycles underlying food production and human nutrition. The evolution of the modern food system around these parameters has imparted distinctive characteristics, setting it apart from other major industrial systems. However, in this chapter we suggest that the inherited configuration of agriculture–industry relations is in a state of flux and identify several leading causes of the restructuring now underway. We focus in particular on the new technological base which is emerging from the application of biotechnologies in agriculture and food processing. In addition, on the demand side, shifts in consumers' perceptions of food are an increasingly important factor of structural change.

In a system so dependent on biological processes, modern biotechnologies have truly revolutionary implications.[1] They confer a *generic* technical capacity to undertake selective genetic intervention in existing forms of life and to create novel life forms. The genetic code can now be manipulated and nature refashioned according to the logic of the market-place. As a result, biotechnologies have suddenly opened up new alternatives and paths of development for *all* major actors in the food system: farmers and input suppliers, primary processors, final food manufacturers and consumers. It is this cross-cutting, polyvalent capacity of biotechnologies which makes

167

them such a potent force in restructuring the food system. Strategic responses by leading actors to the challenges and opportunities these innovations present are defining new modes of integration between agriculture and industry and some, as we shall see, are already discernible.

The transformation of the technological base joins other powerful elements of change at work in the food system. Thus the development of biotechnologies coincides with the crisis of the productionist model and efforts to reform the institutional framework of farm and food policy inherited from the fordist era. If biotechnology R&D is centred heavily on agricultural productivity growth, rather than, say, lower-input, more sustainable systems, this reform process is likely to accelerate. Indeed, as we suggested in Chapter 3, the various pressures released by the 1980s farm crisis – fiscal strains, trade reforms, environmental concern and consumer disquiet – have brought farm policy in the advanced industrial countries to a historic turning point. Biotechnologies are an important element of future scenarios but their application and impact will depend on public regulatory structures and how production and social development policies for rural areas are redefined, with correlate effects on corporate investment decisions and R&D strategies. Similarly, growing public awareness of environmental and food issues is likely to accentuate the shift from a supply-driven to a demand-led food industry, giving consumers and public interest groups more say on such questions as food safety, nutrition, and the ethics of intensive livestock production.

We begin, however, by considering the new modes of integration between agriculture and industry associated with biotechnologies.[2] Since these methods give new, strategic import to ownership and control of genetic resources, this perspective necessarily extends to the Third World, where the majority of the so-called Vavilov centres of bio-diversity are found. With such a wide compass, we are concerned mainly to identify a range of key issues raised by the re-articulation of agriculture–industry relations rather than explore only one or two questions in detail. Our point of departure is to suggest that biotechnologies confront all actors in the food system with new options. The uses to which these innovations are put and the directions taken by the food system will depend on the balance of forces controlling policy and research agendas.

TOWARDS A NEW
TECHNOLOGICAL BASE

The discussion of corn hybridization in Chapter 3 identified the seed as the catalytic element in the convergence of chemical and mechanical technologies after 1930. Industrial property rights to the genetic information incorporated in hybrid varieties received *de facto* protection because the open pollination of seeds harvested from these varieties gave significantly lower yields. This natural or biological protection was reinforced by successful lobbying in the United States to restrict public development of commercial hybrid corn lines and later by the passage of plant breeders' rights of legislation in 1970. The case of hybrid corn has several important lessons for current developments in agricultural biotechnologies, not least the direct link between the industrial appropriation of nature and new trajectories of agro-industrial accumulation. More specific parallels are found in the economic and political pressures to reserve strategic areas for private capital and to create and protect industrial property in genetic material, as we see below.

Modern biotechnologies immeasurably enhance the scope, precision and speed of selective genetic intervention for plant and animal breeding. As the boundaries to industrial appropriation of plant and animal genomes recede, the potential for using genetically manipulated organisms as instruments of production is correspondingly extended. These enormous recent strides towards transforming nature into a productive force under industrial control have built upon advances in the understanding of the molecular structure of biological systems made in the early 1950s, which established the foundations of molecular biology (Perutz 1987). This earlier research includes Watson and Cricks' renowned discovery of the double-helix structure of the DNA molecule, and subsequent work on the genetic code. Commercial applications of the new molecular genetics only became feasible in the early 1970s, however, when a cluster of interrelated innovations created 'the toolkit of the genetic engineer' (Cherfas 1982).

These new methods greatly reduced the time and cost of manipulating genetic material, including the combination of DNA fragments from different organisms, and even different

169

species. By overcoming 'species barriers' to genetic variation, the new genetic engineering techniques of recombinant DNA and cell fusion extend the genetic resource base available for crop and animal improvement beyond the natural gene pool. Plant and animal breeders can now cross the barrier to genetic diversity hitherto presented by sexual fertilization and draw desirable characteristics from other gene pools, engineering 'transgenic' species with specific traits and designed for specific purposes. Biotechnologists have engineered sheep–goat chimeras by mixing embryonic material, and have created transgenic animals with human genes so that commercially valuable human proteins are secreted in their milk. With these technologies, only moral and aesthetic restraints stand in the way of a technicist view of nature 'as a vast organic Lego kit to do with as we please' (Sagoff 1988: 27).

The implications and consequences of this quintessentially industrial engineering approach to genetic selection clearly will depend on the locus of control over biotechnology R&D and the public regulatory framework. Once again, the versatility of agri-biotechnologies is worth emphasizing. In plant breeding, for example, biotechnologies could be deployed to enhance the socio-economic viability of low-input farming systems by endowing crops with biological resistance to disease and pests and, in developing countries, to improve tolerance to resource-poor environments cultivated by the vast majority of rural workers. Alternatively, these new techniques can be used to reinforce existing trends by conferring resistance to proprietary agri-chemicals, designing crops to meet processing requirements, and raising the yields of elite cultivars, including new hybrid varieties patented by multinational seed companies. If we ignore ethical arguments about the means and focus only on ends,

> The same processes . . . can be viewed from diametrically opposed standpoints: as the reversal or reorientation of the inherited industrial model in favour of a more sustainable, environmentally-friendly development, or as the further extension of the dominant industrial paradigm to incorporate the engineering of living organisms.
>
> (Goodman and Wilkinson 1990: 13)

These options, among others, remain open but in practice the directions being taken by R&D in agri-biotechnologies are now becoming more clearly defined. Here we briefly explore the interrelated set of issues raised by increasing concentration in the seed industry, the privatization of agricultural bio-technologies, and the efforts to patent genetically altered life forms. Biotechnologies clearly enhance the significance of the seed as the main vector of technical progress in agriculture. With industrial manipulation now feasible at the cellular and molecular level, specific genetic characteristics, such as herbicide tolerance, can be transferred and expressed in the engineered plant.[3] More than ever, plant genetics hold the key to strategic control over future directions of the agricultural production process and markets for agro-industrial inputs.

CORPORATE APPROPRIATION OF AGRI-BIOTECHNOLOGIES

In recent years, industrial corporations have adopted various strategies to secure a base from which to exploit the emerging opportunities for accumulation created by agri-biotechnologies. The acquisition of seed companies is the common denominator of these efforts, and patterns of R&D organization display greater variety, as we see below. Whatever the motives behind industrial take-overs, the large corporations quickly recognized the commercial logic of combining genetic research with conventional plant-breeding capacity and international marketing networks. Realization that plant biotechnologies are marketed through the seed similarly has prompted newly established genetic research firms, such as Calgene and Plant Genetics, to acquire seed companies. Successive waves of acquisition are well documented in the literature and it is estimated that major chemical, pharmaceutical and oil companies have taken over more than 120 seed firms since the mid-1960s (Kenney *et al.* 1983). Industry observers suggest that the current wave of take-overs is driven by the need 'to consolidate market share in selected species, in anticipation of products from biotechnology' and by recognition that 'seed and plant biotechnology are pivotal for proprietary food and animal feed' (Kidd 1987: 7).

ICI has been one of the most active multinational players in the current round of take-overs, acquiring Garst Seeds, the third largest US corn seed producer, in 1985 and a leading UK seed firm, Sinclair McGill, in 1986. Its acquisition of the Société Européenne de Semences (SES) in 1987 took ICI into the ranks of the top world seed producers, where it joined other multinational agri-chemical firms, such as Monsanto, Ciba-Geigy, Shell and Sandoz. ICI also was an unsuccessful bidder for the famous Plant Breeding Institute at Cambridge, which was purchased by Unilever following its privatization by the British government in 1987. The restructuring and internationalization of the world seed industry seems set to continue at a rapid pace since the top dozen companies, all multinationals, at present account for only 12 per cent of the estimated US$26 billion turnover in this market (Kidd 1987). This observer foresees a highly concentrated structure by the year 2000, when 'between 10–20 multinational-owned seed and plant biotechnology companies will dominate the most profitable, research-intensive segments of the seed trade. (Kidd 1987: 8).

The accelerating global reorganization of the seed industry has been paralleled by rising levels of concentration in agri-chemicals. Thus in 1985 DuPont purchased the US agri-chemical interests of Shell, while Rhone-Poulenc took over the activities of Union Carbide. ICI again made significant progress in world rankings following its take-over of Stauffer Chemicals in June 1987, which took it to third place in world agri-chemical sales behind Bayer and Ciba-Geigy. The strategic complementarities which these rounds of acquisitions in the seed and agri-chemical industries are intended to exploit have been widely recognized. 'For ICI and its rivals, the aim is to create an agricultural package reaching from fertilizers through pesticides to the plant itself, which can be tailor-made through genetic manipulation to fit the maker's system and no one else's' (Lex, *Financial Times*, 13 June 1987). Research by Calgene to develop varieties resistant to the Monsanto herbicide, 'Roundup', offers a well-known illustration of this direction of corporate strategy to protect and extend markets for proprietary agri-chemicals, including conventional inorganic compounds. Although integrated proprietary input packages may not be an immediate prospect, these developments reveal the restructuring potential of biotechnologies and explain the recent rush to establish a general competence in this new field, particularly by those firms whose core businesses are directly affected.

The strategies employed by multinational corporations to gain access to biotechnologies have been strongly influenced by the emergence of molecular genetics in university research centres. These academic origins were clearly reflected in the structure of the early biotechnology industry which developed in the early 1970s following the introduction of the genetic engineer's 'toolkit'. In these early days, the majority of firms, especially in the United States, were small specialized research companies, set up in many cases by university researchers with the support of private venture capital. However, difficulties in funding R&D expenditures soon proved to be the Achilles' heel of these small companies. Commercial inexperience and the lack of revenue-yielding products to offset high R&D costs, exacerbated by long, unpredictable lags between laboratory trials and regulatory approval, have made start-up firms vulnerable to larger corporations, whether through take-overs or other links. These linkages variously include equity participation, joint ventures, limited R&D partnerships and the funding of venture capital groups which invest in biotechnology. Other modes of entry comprise university contract research and more permanent corporate–university links, as well as the development of in-house R&D capacity. Some companies, such as Hoechst, ICI and Monsanto, have implemented these strategies concurrently.

These trends exemplify the internationalization of biotechnology R&D and its growing concentration in the hands of the oligopolistic multinationals that dominate the main potential user industries – agriculture, chemicals, pharmaceuticals, energy and food.

The evidence that the logic of biotechnological innovation . . . is shifting the centre of gravity overwhelmingly in favour of multinational and 'Fortune 500' corporations is inescapable . . . The large corporations, using their R&D experience, vast resources and flexible funding methods, now dominate commercial technology and, increasingly, the direction of fundamental research.

(Goodman *et al*. 1987: 110)

The 'Black Monday' stock market crash of October 1987 further exposed the structural weaknesses of small genetic research firms in terms of R&D funding, limited manufacturing facilities, and marketing networks.[4] With tighter capital markets, these firms have turned with growing frequency to 'strategic alliances' to finance R&D expenditures. European and Japanese corporations have been particularly quick to exploit this mechanism to gain access to US biotechnology R&D via contract research and the acquisition of marketing franchises (Burrill 1989; Dibner 1989). These clientelistic alliances compromise the independence of small genetic research firms, which effectively transfer some control over R&D policy to their multinational customers.

Although these patterns are found in all the main areas of biotechnology, certain structural features suggest that they are particularly likely to reinforce concentration tendencies in agri-biotechnology. Thus recent strategic alliances in this sector, such as those betwen Ciba-Geigy and Agri-Diagnostics, Rhone-Poulenc and Calgene, and Roussel Uclaf and Calgene, are symptomatic of the emerging leadership of Western European companies (Walker 1989). With some notable exceptions, including Monsanto and Du Pont, the US agri-biotechnology sector is characterized by many small specialist start-up companies, whereas European R&D is strongly dominated by large, diversified multinationals for whom innovation in the seeds/chemicals nexus is vital to competitiveness in their core business sectors. The movement of leadership in agri-biotechnology towards Western Europe therefore is likely to be accompanied by more restrictive access to R&D, further accentuating its already pronounced corporate identity. As Walker observes, in contrast to the present situation in the United States,

> external companies have limited access to European expertise, *as most of it is jealously guarded by its corporate owners*. These corporations represent powerful players. They include Bayer, Ciba-Geigy, and ICI, the world's three largest agri-chemical companies, . . . large agribusiness corporations such as Unilever and Nestlé, and major chemical companies such as Rhone-Poulenc, Montedison, Shell, Hoechst and Enichem.
>
> (Walker 1989: 123, emphasis added)

As we have seen, these same corporations are taking an active role in restructuring the seed industry.

THE PRIVATIZATION OF AGRI-GENETICS RESEARCH

These recent developments indicate that the nascent agri-biotechnology industry already is assuming the oligopolistic configuration characteristic of the core business sectors of its leading multinational protagonists. Furthermore, complementary changes in the institutional structure of agricultural research, provoked by international technological competition and neo-liberal policies, are strengthening corporate domination of agri-biotechnology. The dividing line between public and private research is being redrawn, with critical implications for the control of germplasm and institutional channels of technology transfer, as we shall see below.

The main thrust of this restructuring is to direct the public sector towards basic research and away from its traditional activities in applied plant breeding and the release of new crop varieties into the public domain. The concerted acquisition of seed companies by corporations with agri-biotechnolgy interests fits neatly into the new schema of privatizing 'near-market' R&D on plant improvement. This changing division of labour in agricultural research is accompanied by corporate pressures for stronger patent protection in agri-biotechnologies and the introduction of regulatory systems which routinely permit the release of genetically manipulated organisms into the environment.

Two complementary stages can be discerned in the changing public–private sector relations in agri-biotechnology research. Initially, private corporations scrambled to expand their R&D capability and secure a 'window' on the new field of molecular genetics technologies. We have briefly noted the various strategies corporations employed, often concurrently, for this purpose. By the mid-1980s, private corporations were well-positioned to contribute to developments on the frontiers of agri-biotechnology research and to control commercial applications through their ownership of seed companies. With this foundation, corporate interests and public policymakers turned their attention to the implementation of a more 'rational' division of labour in research. Although cloaked in the powerful

175

discourse of 'national competitiveness', essentially this involves putting fundamental research in the public sector at the service of the applied, 'near-market' development work undertaken by industry.

Given the origins of biotechnologies, university–industry relations provide an early and continuing indication of the emerging division of labour (Kenney 1989). Examining the US Land Grant University (LGU) system, Buttel (1986a) notes that these links were traditionally confined largely to private grants for development-oriented research and the distribution by the LGUs of public domain products, such as finished crop varieties. By the mid-1980s, however,

> these relationships had become closer, more formalized; the industry share of experiment station budgets has risen. Industry grants are for larger amounts, are more likely to involve proprietary considerations, and are more likely to involve fundamental rather than highly applied or developmental research. Further, LGUs have largely moved away from releasing . . . finished crop varieties.
>
> (Buttel 1986a: 126)

University–industry contract arrangements thus not only direct the research agenda by pre-empting public research capacity but also spread the mantle of trade secrecy over its output, impeding the free flow of information and biological materials around the public research network. This encroachment, in turn, inhibits the transfer of technology, including the exchange of germplasm, between national and international institutions engaged in plant improvement work.[5] These trends are destroying traditions of openness and collaboration which have long characterized the research culture of public plant breeding (Kenney 1989).

We have suggested that the restructuring of public agricultural research in the 1990s has responded directly to the changing interests and R&D capability of the corporate sector, overlaid by the chauvinistic rhetoric of global technological competition. In Europe, the immediate objective of public and corporate policymakers was to 'catch up' with the levels of commercial biotechnology R&D found in the United States and Japan. This aim was pursued in the early 1980s by actively promoting the formation of research 'clubs' and other consortia to accelerate the transfer of biotechnology from the universities and public research institutes into industry. Inevitably, therefore, commercial considerations assumed greater importance in determining directions of research. In the United Kingdom, closer industrial collaboration and greater commercial 'relevance' effectively were forced on the public agricultural research system by cuts in funding. International technological rivalry, though now in the context of global trading blocs, also has been the spur used to rally support for the joint research programmes launched by the Biotechnology Concertation Unit in the European Community.

Collaborative ventures were supported as an effective means of transforming basic research into commercial technologies and marketable products. An alternative to the precompetitive research 'club' approach was adopted by the Agricultural and Food Research Council (AFRC) in 1983, when it established the Agricultural Genetics Company (AGC) to commercialize public sector research. AGC, whose shareholders include Ciba-Geigy, Ultramar, Eli Lilly and venture capital groups, was given first rights to exploit plant biotechnology research undertaken at six AFRC centres. More recently, AGC has acquired rights to some of the AFRC's plant-breeding research. For example, 'AGC has exclusive commercial access to what some plant scientists regard as the world's best collection of *Rhizobium* at the AFRC's Rothamsted Experimental Station in Hertfordshire' (Connor 1988: 49).[6]

As corporations have developed greater R&D capability, combining in-house facilities, equity in start-up biotechnology research firms, and access to the public science base through 'club' and other contractual mechanisms, government funding policy has decreed a further change in the direction of public agricultural research. Although the closer university–industry research links of the early 1980s are seen as a welcome development, the emphasis on commercial relevance and exploitability in public research is now being modified in order to redraw the demarcation line between the public and private sectors. Increasingly, the rightful domain of the public system is defined as basic research, with its greater risks, long-term 'blue sky' characteristics, and uncertain gestation. Thus the then British Minister responsible for Science, Mr Robert Jackson, stated in May 1989 that 'we will now be increasingly withdrawing from near-market research. But we will continue to emphasise our commitment to the Government's own responsibility for supporting *basic* research' (AFRC News, October 1989: 4, emphasis in original).

In the United Kingdom, this re-orientation was announced by two White Papers published in 1987–8 and has led to the privatization of 'near-market' activities, as in the case of the dismemberment and sale to Unilever in September 1987 of the AFRC's Plant Breeding Institute (PBI) and its sister organization, the National Seed Development Organization (Webster 1989).

> the PBI housed the AFRC's core set of plant molecular biologists whose primary objective was to incorporate their plant biotechnology techniques within the more traditional techniques of the breeders, 'on behalf of British agriculture'. Employing over 250 staff, the PBI was regarded as the 'jewel in the AFRC's crown'.
>
> (Webster 1989: 225)

This 'near-market' policy has resulted in major cuts in government R&D expenditure,[7] the closure of public horticultural centres and, perhaps most contentiously, of research institutes working on food safety despite recent outbreaks of salmonella, listeria and bovine spongiform encephalopathy. Popular sentiment against this policy has been stirred by the decision in December 1989 to close the National Fruit Trials station at Brogdale, which maintains the national genetic fruit bank and has the most comprehensive apple collection in the world.

The definition of 'near-market' research is equivocal and it tends to be identified in practice with industrial funding. Of course, since corporations may have interests in selected areas of basic research, this criterion can lead to under-funding of the public system and even greater dependence on industrial collaboration.[8] In general, however, the 'near-market' approach is intended to ensure that public monies are used to finance research activities that industry will not support. The public system is being cast as the handmaiden to industry in the international competition to develop commercial agri-biotechnologies. This is the significance of the new division of labour in agri-biotechnology research, whose framework began to emerge in the United Kingdom in the late 1980s.

178

Such pressures were articulated at an earlier stage in the United States, crystallizing in the recommendation of the Winrock Report (Rockefeller Foundation 1982) that the public Agricultural Research Service and the Land Grant University (LGU) system should devote greater attention and resources to fundamental research. This report, couched in terms of global competition in 'cutting edge biological science', 'provides a template for the reshaping of public agricultural research along lines more responsive to the changing needs of capital' (Kloppenburg 1988a: 235). The policy changes advocated by the Winrock Report are now being implemented through federal funding programmes. These priorities are also being applied to the LGU system by increasing the share of research monies allocated via the USDA's Competitive Research Grant system at the expense of general institutional or 'formula' funding. Kloppenburg is convinced that, under the new division of labour, 'Public agencies will move increasingly towards basic research in which biotechnology will be used to evaluate and improve germplasm in what is known as the pre-breeding phase of plant improvement' (Kloppenburg 1988a: 241).

The new role assigned to the public sector in agri-biotechnology research acquires particular significance in the context of a greatly expanded public research effort. These new priorities are endorsed by the US National Research Council report, *Agricultural Biotechnology: Strategies for National Competitiveness*, which calls for federal funding in this field to reach US$500 million annually by 1990 (Moses *et al.* 1988). This level is more than three times private research expenditure of $150 million in 1985, which at that time represented nearly two-thirds of *total* outlays on agriculturally-related biotechnology. In the early 1980s, this glaring disparity had aroused fears in the United States that corporate power alone would be sufficient to gain control over the research agenda, with the public system conceding by default. Thus Kenney *et al.* warned that 'the high capital-intensity of biotechnology-related plant breeding will render public breeding technologically antiquated *vis-à-vis* its more adequately funded counterpart in the private sector' (Kenney *et al.* 1983: 486)

With hindsight, this warning, although apposite, misread the changes underway; private leadership is being conferred not by default but by design, supported by expanded complementary public funding. Looking ahead to the 1990s, the balance of control over agri-biotechnology is being swung towards the corporate sector by policy decisions deliberately intended to weaken the bipolar institutional framework characteristic of conventional plant breeding in the United States. In agri-biotechnology, public resources will increasingly be concentrated in generic, precompetitive research, relinquishing 'near-market', applied R&D activities to private plant breeders (Buttel and Busch 1988).

The public agricultural research system in the United States has historically been distinguished by the link between basic research and decentralized, applied development work to produce finished, locally-adapted varieties. In the case of agri-biotechnology, this link is now being severed, compromising the role of the public system in future plant development work. As Buttel observes, 'Public agricultural biotechnology R&D policy is currently being formulated in a virtual vacuum . . . mainly by recourse to ideological "black boxes", such as "national competitiveness", "university–industry cooperation", and so on' (Buttel 1989b: 21). In the United States, as in Western Europe and elsewhere, the imagery of global markets and 'high technology' has combined powerfully with neo-liberal policies to devolve control over the research agenda and deployment of agri-biotechnology to the corporate, that is to say multinational, sector. These trends will have an important influence on the emerging structure of the global food system, including the possible realignment between North and South in international agricultural trade.

PATENTING THE BIOSPHERE

Pressures to recognize industrial property rights to genetically modified organisms and new life forms have grown in tandem with corporate R&D investment in biotechnologies. In agri-biotechnologies, mounting efforts to strengthen intellectual property rights over living organisms, allied to the trade secrecy that already envelops much research, lend great urgency to debates on patenting, modes of technology transfer and, most fundamentally of all, the ownership of genetic resources. These debates, with their moral and ethical dimensions, are increasingly being cast in a framework of North–South confrontation.

These tensions are exacerbated by growing fears of the negative consequences these new technologies may have on Third World countries, particularly those whose technological capability is more limited.

Demands for intellectual property protection in agri-biotechnologies arise in the highly specific context created by the trends in technological change, industrial restructuring, and institutional reorganization we have discussed above. Yet efforts to establish intellectual property rights in plant breeding have a long history, oscillating between advocates of industrial patent protection for living organisms and proponents of a separate legislative framework of plant breeders' rights or PBRs (Juma 1989). This separation was recognized by the International Convention for the Protection of New Plant Varieties (UPOV), signed in 1961, mainly by industrial countries, which states that signatories may not resort to both patents and PBRs. This provision does not extend to countries that had allowed both forms of protection before joining UPOV, notably the United States. However, the accelerating flow of innovations in biotechnology involving genetically modified living organisms which are used as products and processes in a wide variety of sectors has made the separation between industrial protection and agricultural protection increasingly artificial. Indeed, the restructuring process set in train by these technologies is blurring even the more basic distinction between 'industry' and 'agriculture', as we argue below.

As innovative activity has intensified, a series of legal decisions in the United States has extended patent protection to genetically engineered living organisms, displacing the 'product of nature' principle which had previously dominated this area of patent law (Juma 1989).[9] In the landmark case of *Diamond v. Chakrabarty* in 1980, the US Supreme Court narrowly ruled that 'a live human made micro-organism is patentable subject matter' (quoted in Kloppenburg 1988a). However, whether this decision included plants remained uncertain due to the protection already provided by the 1930 Plant Patent Act covering asexually reproduced varieties and the 1970 Plant Variety Protection Act (PVPA). This uncertainty, which effectively maintained the *de facto* separation between industrial patent protection and PBRs, was dispelled by the decision of the US Board of Patent Appeals and

Interferences in *Ex parte Hibberd* in 1985. This decision grants protection to plants under utility patent protection, breaking the long tradition of separate property rights by allowing 'breeders to choose among the several statutes for the best form of protection' (Kloppenburg 1988a: 263). Following its decision in April 1987 in *Ex parte Allen*, the US Patent and Trademarks Office 'now allows for the patenting of genetically-manipulated non-human multicellular living organisms, including animals' (Juma 1989: 164). Parallel efforts to strengthen intellectual property rights to living organisms are being made in Europe and in October 1988 the European Commission issued its draft *Proposal for a Council Directive on the Legal Protection of Biological Inventions*. One observer suggests that this draft 'gives the most expansive reasonable interpretation of the patent right' (Barton 1989: 264–5).[10]

These recent developments in property rights legislation establish a much more restrictive regime of protection in plant and animal breeding, with far-reaching implications for industrial structure, farmers' access to varietal improvement, and agriculture in Third World countries. As Kloppenburg indicates, claims for a new variety under the PVPA treated the plant as an indivisible whole, whereas 'Utility patents may encompass claims not only to multiple varieties but also to the individual components of those varieties: DNA sequences, genes, cells, tissue cultures, seed, and specific plant parts, as well as the entire plant.' The *Hibberd* decision consequently opens the door to 'the licensing of particular components – e.g., a gene for herbicide tolerance – use by third parties' (Kloppenburg 1988a: 263–4).

On a wider canvas, the *Hibberd* decision poses a direct threat to farmers' age-old practice of saving harvested seed for replanting, since this is likely to infringe the utility patent statute. If this proves to be the case, the intense biotechnology R&D activity in the principal commercial crops suggests that farmers will become more dependent that ever on seed firms and their parent agri-chemical corporations. This prospect of enforcing property rights to genetically modified seed among farmers gives added significance to the secondary and subordinate role assigned to the public sector in the current institutional reorganization of plant breeding.[11]

This marginalization will accelerate even further if the courts uphold the increasing number of claims for *process* patents on the techniques of recombinant DNA and cell fusion which are pushing back the frontiers of plant improvement work. The gradual disengagement of the public sector from varietal development and the tightening web of patent protection in agri-biotechnologies reveal the potential scope for monopoly rents in the seed industry.

Against the background of global technology rivalry, including growing competition from the newly-industrializing countries, national geo-political interests and corporate strategies have united behind efforts to strengthen intellectual property rights at the international level. The United States and other advanced industrial countries canvassed vigorously for a new code of intellectual property rights during the recent Uruguay Round of GATT trade negotiations. Concurrently, discussions to harmonize international patent legislation relating to biotechnology have been held under the auspices of the World Intellectual Property Organization (WIPO). Further talks on harmonization between the European Patent Office (EPO), the US Patent and Trademark Office and the Japanese Patent Office are believed to be closer to agreement than the WIPO negotiations. These initiatives add to the pressures on Third World countries to recognize patents on life forms, including those incorporating genetic material which may well have originated in these countries. In the meantime, the United States is using the threat of trade sanctions to persuade countries to strengthen their intellectual property systems so that US patents can be enforced.[12] Amendments to the 1988 Trade Act enable the US authorities to draw up an intellectual property 'Priority Watch List' for this purpose (Barton 1989).

PLANT GENETIC RESOURCES: COMMON HERITAGE OR PRIVATE PROPERTY?

This aggressively competitive climate of privatized agri-biotechnology R&D, 'creeping secrecy' and concerted efforts to patent life forms has given renewed prominence to several key issues which first came to the fore in debates about the Green Revolution. These include ownership of genetic resources, Third World access to germplasm collections, genetic erosion and other environmental impacts of intensive, science-based agriculture. The salience of these issues reflects the virtually unlimited possibilities unlocked by biotechnologies of transferring genetic material across species barriers. These new methods potentially link all existing gene pools, permitting plant breeders to select genes from other plant and animal species and microbiota. However, the Vavilov centres of genetic diversity, and consequently the principal repositories of the raw materials of biotechnology, are found mainly in the Third World. There has been conflict over the exploitation of these resources since the 1970s, and the pressures to broaden intellectual property rights to include patentable life forms suggests that these disputes are entering a new, more intense phase.

The Vavilov centres have provided a continuing flow of germplasm to the metropolitan countries of the North since the early colonial period.[13] More recently, as Juma reminds us, 'The rise of the United States as a global agricultural power was as a result of the persistent efforts to introduce new genetic material and the related technology into the economic system' (Juma 1989: 3). The systematic, heavily asymmetrical transfer of germplasm to the North was institutionalized within the network of IARCs established in the 1960s and 1970s, and reached its apotheosis in 1974 when the International Board of Plant Genetic Resources (IBPGR) was created. This new institution was founded as an integral element of the CGIAR system with a mandate to co-ordinate the collection and conservation of germplasm in an international network of gene banks.[14]

As we saw in Chapter 4, the IARCs were the main conduit for the transfer of new plant-breeding techniques and the diffusion of HYVs and complementary chemical technologies in Third World agricultures. By a cruel paradox, though hardly unforeseen, the very success of this mission is concomitantly eliminating the genetic diversity on which all agricultural production ultimately depends. These elite varieties, custom-built for fertile, protected environments, are displacing the indigenous land races and primitive cultivars whose genetic material is incorporated in their parent lines and whose extinction would be an inestimable loss for humankind.

Genetic erosion in the Third World and the narrow genetic base of modern agriculture were increasingly recognized as potentially dangerous consequences of monocultural specialization and selective plant-breeding programmes. However, the greater vulnerability of modern elite varieties to disease and pests was painfully demonstrated in 1970, when southern corn-leaf blight (*H,maysis*) caused an estimated 15 per cent loss in US corn production. This epidemic focused attention on the genetic uniformity of most commercial crops grown in the United States, and the correspondingly vital importance of genetic resources and their conservation. 'As the gene pool for a species is drained by genetic erosion, it becomes more difficult to find characteristics to combat the appearance of disease and pest epidemics that challenge the genetically vulnerable commercial cultivars' (Kloppenburg 1988a: 162). These events in the heartland of US agriculture, which perceptibly threatened national food security, spurred efforts to organize an international system to conserve plant genetic resources, culminating in the foundation of the IBPGR.

The decision to establish the IBPGR within the CGIAR network, so closely identified with the advanced industrial countries, rather than in the United Nations system, which would afford member countries full and equal access, was controversial and continues to foster doubts about its legitimacy (Mooney 1983; Kloppenburg 1988b; Juma 1989). These have been nurtured by the related decision to designate existing gene banks as global 'base collections', since two-thirds of these are located in the North and the CGIAR international research centres. These institutional arrangements have constantly added fuel to disputes about the ownership of *ex situ* germ-plasm collections and the legal standing of *in situ* genetic resources. On the first issue, Juma notes that

> The question of ownership not only relates to access by countries of origin but also deals with the legal status of agreements entered between the IBPGR with the con-serving banks, especially national facilities which are subject to national law.
>
> (Juma 1989: 90)

Third World countries not surprisingly are alarmed that germ-plasm freely collected on 'common heritage' principles under IBPGR auspices may, in fact, constitute national property, with access therefore subject to political decision.

The legal standing of *in situ* plant genetic resources provides the real focal point of these international 'seed wars', however. This issue crystallizes the ambivalent treatment of plant germplasm in the North and the growing vulnerability of the South as intellectual property systems are adapted to cover agribiotechnology innovations. The 'gene-poor' North resolutely defends the convenient distinction between plant germplasm as a 'common heritage' resource and as private property. 'Unimproved' or 'primitive' germplasm is considered the common heritage of humankind, and therefore is a public good, freely collectable in the Third World through the CGIAR network and freely available to plant breeders. Plant germplasm becomes private property only when embodied in improved 'elite' varieties and commercial breeding lines. Third World germplasm not only is commoditized as seed and intellectual property but there also is no guarantee of *reciprocal* freedom of access to the breeding lines incorporating this material nor compensation of any kind. Recent patent legislation, epitomized by the *Hibberd* decision which grants protection both to genetically engineered *whole* plants and single genes, gene complexes, and highly specific characteristics, has sharpened awareness of these contradictions and the strategic significance of genetic resources.

This rapidly evolving situation has given renewed urgency to Third World demands that the principles of common heritage and free exchange be extended to all types of germplasm. This proposition is rejected out of hand by the advanced industrial countries which, as we have seen, are pressing insistently for more comprehensive intellectual property rights in the current GATT negotiations and other international fora. Third World proposals have been articulated through the FAO, leading in 1983 to the creation of a Global System of Plant Genetic Resources. This as yet skeletal system consists of a basic legal framework, the International Undertaking on Plant Genetic Resources, an inter-governmental forum, the Commission on Plant Genetic Resources (CPGR) and the International Fund, a financial mechanism established in 1987 to support the conservation and sustainable utilization of plant germplasm.

The International Undertaking, reduced to a voluntary agreement rather than a legally binding convention after concerted pressure by the advanced industrial countries, embraces the common heritage principle for all plant genetic resources, including free access to proprietary breeding lines and finished varieties, and mandates the development of an international network of base collections in gene banks under the jurisdiction of the FAO. Given the sustained hostility of the advanced industrial countries to these initiatives, the recent decision by the Board of Trustees of the IBPGR in principle to move from FAO headquarters in Rome to Copenhagen suggests that a dual system of genetic conservation may evolve in the 1990s (Esquinas-Alcazar 1989). At its Second Session in 1987, the CPGR recognized plant breeders' rights as provided for under UPOV but proposed that the concept of farmers' rights be given equal legal standing.[15] This concept was endorsed at the CPGR's Third Session in April 1989, which supported farmers' rights to compensation for the use of their genetic resources in conventional plant breeding and agri-biotechnologies. In advocating farmers' rights, the FAO draft resolution considers that 'in the history of mankind, unnumbered generations of farmers have conserved, improved and made available plant genetic resources', and that 'farmers, especially those in developing countries, should benefit fully from the improved and increased use of the natural resources they have preserved' (FAO 1989b: 13).

In view of the implacable opposition and geo-political weight of the advanced industrial countries, it is difficult to foresee common heritage principles usurping intellectual property rights in plant breeding. Indeed, this opposition may compel Third World countries to pursue alternative paths, possibly using the concept of farmers' rights to assert national sovereignty over plant genetic resources.[16] This step would encourage the formation of a market in plant genetic resources, which would create pecuniary incentives for North and South alike to promote biodiversity. National-property initiatives similarly would stimulate recourse to multilateral agreements to resolve North–South conflicts, as well as problems arising from unequal plant genetic resource endowments among developing countries. The principle of national property might also provide a more effective basis from which to negotiate compensation for possible negative impacts of biotechnologies on Third World economies and agricultural export markets.

RESTRUCTURING NORTH–SOUTH RELATIONS

These possibilities emphasize the crucial significance of the changing political economy of agricultural research in restructuring the global food system. Several issues demand our attention but they have common origins in the locus of control over the research agenda in agri-biotechnologies, and therefore over their deployment. Thus there is legitimate concern that rising privatization has biased this agenda heavily towards the economic priorities of the advanced industrial countries. Not surprisingly, commercial agri-biotechnology research, which is profit-orientated and therefore market-driven, has neglected the so-called 'minor crops' produced in tropical Third World countries by small farmers for local consumption. However, this neglect is compounded by fears that growing corporate control and more restrictive intellectual property systems will inhibit technology transfer. Such obstacles threaten to hinder efforts by Third World countries to develop agri-biotechnologies appropriate to tropical agro-ecosystems and indigenous cultural practices and which meet the needs of poor family farmers.

This deliberate neglect would represent a tragic lost opportunity, but Third World agricultures are already vulnerable to developments arising from the Northern bias of plant biotechnologies and recent advances in industrial fermentation and industrial tissue culture. These impacts, integral elements of the process designated earlier as substitutionism, are central to the restructuring of North–South relations in the agri-food system. Plant and fermentation biotechnologies can be competitive or complementary. For example, single cell proteins (SCP) cultured on non-agricultural substrates can displace livestock products or, alternatively, the nutritional/chemical content of crops can be genetically 'engineered' to meet industrial processing requirements. The complex possibilities for substitution created by biotechnologies between industrial and agricultural products, and between different agricultural commodities, obviously constitute powerful forces for change in the agriculture of advanced industrial countries (Goodman *et al.* 1987).

Similar forces, but far more disruptive to rural livelihoods, are at work in redefining the role of Third World agricultures in world trade, with potentially disastrous effects on rural production and welfare, particularly in highly specialized, monocultural exporting countries. Such adverse consequences will arise directly from corporate R&D decisions, producer strategies, and agricultural policymaking in the North. Substitution of Third World commodities can occur as the result of displacement through factory production, by enhancing the comparative advantage of temperate crops, and innovations which increase the interchangeability and competition between crops as industrial inputs. The first category includes the application of plant tissue culture in the factory production of a wide range of food flavours, colours, oils, spices, and non-nutritive sweeteners such as monellin and thaumatin derived from West African plants (Sahai et al. 1986).[17] Tissue culture production of vanilla currently threatens the livelihood of 70,000 producers in Madagascar (FAO 1989a). More sweeping impacts on spatial and climatic barriers to production will follow if efforts to produce tea, coffee and cocoa by tissue culture or in vitro methods are successful. As Busch et al. remark, 'These flavour components would be identical biochemically to the compounds naturally found in these products; hence, they would not be artificial in the sense now currently understood, but would be true equivalents' (Busch et al. 1989: 125).

Spatial and climatic determinants of production location and comparative advantage also are challenged by the combination of industrial country policies to stimulate agricultural import substitution, as in the case of oil seeds (rape seed, sunflower, etc.) in Western Europe, and plant biotechnologies to improve yields and processing qualities of temperate crops. Concomitantly, biotechnologies applied to bacterial fermentation methods are reducing the production costs of animal feed additives, such as 'essential' amino acids, putting new competitive pressures on higher-priced feeds. These additives can be used to enrich cheaper, low-protein feed, such as corn meal, threatening exports of soya beans and manioc from Brazil, Argentina and Thailand. Animal biotechnologies to increase feed conversion ratios and adapt livestock to low-quality feed, such as cellulose in the case of pigs, work in the same direction. A further example with direct consequences for Third World producers involves research to reproduce the quality of palm oil by the enzymatic treatment of lower-grade oils. The counter-strategy to enhance the competitiveness of palm oil by using tissue culture to clone higher-yielding, disease-free varieties of oil palm is being pursued by Unilever, reportedly with very mixed results.

The impact of biotechnologies in accentuating the flexible use of agricultural crops and intensifying competition between traditional product complexes is graphically illustrated by the production of high fructose corn syrups (HFCS) in the corn wet milling industry. Cost-reducing innovations in genetic engineering and immobilized enzyme technology, combined with high US support prices for sugar, have led to the large-scale substitution of HFCS for cane sugar in industrial food processing and soft drinks production in the United States and Japan.[18] This has displaced exports from leading Third World suppliers, such as the Philippines, creating unemployment estimated in the millions, and depressed world prices.[19] With these new industrial biotechnologies, cane and beet sugar join a wider category of 'sweeteners' derived interchangeably from a variety of starch sources, including corn, wheat, manioc, sorghum and potatoes. This new flexibility has prompted two leading sugar multinationals – Tate & Lyle and Ferruzzi – to diversify into sweeteners via take-overs of firms specializing in HFCS and other products of starch chemistry based on agricultural feedstocks.[20] These developments presage new industry–agriculture relations as plant *and* industrial biotechnologies forge a new alliance betweeen a *polyvalent* agriculture integrated with a diversified processing sector supplying inputs for food, feed, fuel, chemicals and pharmaceutical products (Goodman and Wilkinson 1990).

In this respect, HFCS and other cereal sweeteners are the forerunners of a much broader trend impelled by advances in industrial bio-processing, which reduces field crops to their generic constituents (proteins, starches, sugars, oils) for downstream food and non-food industries. Biotechnologies thus promise to *implode* the traditional organization of the agri-food system around specialized commodity chains by diversifying product markets and pitting commodity complexes against each other. 'As this process of implosion removes long-standing barriers, the agri-food system gradually will merge with the chemical and pharmaceutical industries to form a "bio-industrial processing complex"' (Goodman and Wilkinson 1990: 26). Within this complex, downstream food and non-food products will share common inputs and a common technological base.

The effect of this broad substitutionist tendency is to *trivialize* agricultural commodities, undermining their specific, 'customized' markets by transforming them into polyvalent but generic industrial raw materials. Although this tendency overrides a simple North–South polarization, Third World countries risk being consigned to the status of producers of low value-added biomass materials for generic, intensely competitive markets. This is likely to reinforce downward pressures on their terms of trade with adverse consequences for long-run growth and welfare. In Third World agricultural economies, future prospects will depend significantly on their capacity to utilize plant and industrial biotechnologies to maintain the competitiveness of export crops and diversify into higher value-added products. In turn, success in establishing bio-processing industries based on local crops also holds the key to the internal transformation of Third World food systems and greater technological independence (Wilkinson 1987).

This discussion emphasizes the global reach of biotechnologies in restructuring the agri-food system. Equally, it makes clear that these are *enabling* methods with varied and conflicting applications, which consequently generate new competitive pressures, structural tensions and countervailing tendencies.

> An important corollary is that the relative exclusiveness of different eco-systems, which often has imparted geographical specificity to agro-industrial organization, also is threatened by biotechnologies. The current reorganization of agriculture–industry relations therefore also implies a global geographical realignment.
>
> (Goodman and Wilkinson 1990: 39)

TECHNOLOGY TRANSFER

These considerations provide a strong rationale for Third World countries to acquire the capability to develop biotechnologies adapted to their needs and priorities. Further justification is the possible contribution of biotechnologies in enhancing the productivity of low-input farming systems and the sustainability of marginal agro-ecosystems. Of course, this abstracts from the role of national and international power structures in determining applications of new technologies, perpetuating the false claim that technological impacts can be evaluated separately from questions of access, class structure and the distribution of income and wealth. The lessons of the Green Revolution and the Third World food crisis teach otherwise.

In the case of plant biotechnologies, we have stressed that the social organization of R&D is being transformed by the concentration of applied crop improvement work in private hands. The implications of Northern R&D bias and the vertical integration of multinational agro-industrial corporations are neatly drawn by Silva, who argues that 'we are more likely to see herbicide not weed-tolerant, pesticide not pest-resistant, nitrogen-responsive not nitrogen-fixing, and water-responsive not drought-tolerant varieties' (Silva 1988: 64). In addition, as we have seen, intellectual property systems are becoming more restrictive, extending to individual genes and highly specific genetic traits.[21] Indeed, recent patent decisions 'might be interpreted to mean that a patent holder could prevent others from undertaking competing research, even though different genes are involved and other techniques used' (FAO 1989a: 6). These trends indicate that earlier organizational forms of North–South technology transfer in plant genetics, namely, the IARCs and the CGIAR system, will be inadequate. Even so, these institutions represented the geo-political interests of the North as agents of the Green Revolution and agricultural modernization strategies, which disseminated unequalizing, capital- and energy-intensive technologies.[22] Higher aggregate production has been accompanied by widespread malnutrition, greater landlessness and increased rural income inequality. Mexico provides the *locus classicus*.

In the context of plant biotechnologies, these institutions are again likely to follow a research agenda responsive to external interests and heavily dependent on imported technology. This is the implication of CGIAR statements that the IARCs will be 'users' rather than 'originators' of the new technology (Deo and Swanson 1987). This restricted role tacitly acknowledges that commercial interests will be uppermost in determining the transfer of the products and processes of agri-biotechnologies to the Third World and the exchange of germplasm. As Buttel (1986b) and many others have observed, the centre of gravity in relations between private biotechnology-seed companies and public agricultural research centres is likely to move away from co-operation and exchange towards competition and secrecy.

This changing climate was influential in the strong Third World support for the UNIDO initiative to establish the International Centre for Genetic Engineering and Biotechnology (ICGEB) in 1983. This later was divided into two centres, one in Trieste and the second in New Delhi. However, opposition from advanced industrial countries, notably the United States and Japan, has restricted its effectiveness as a mechanism to accelerate the international transfer of biotechnologies. Thus, in 1987, 'UNIDO withdrew a good part of its commitment to IGCEB and biotechnology in general . . . [which] meets the interests of certain industrialized countries with well-developed biotechnology capacities, which had opposed IGCEB from the very beginning' (Fowler et al. 1988: 324–5). Higher hopes for building biotechnolgy capability are vested in the UNEP network of Microbiological Resource Centres (MIRCENS), possibly in future collaboration with the ICGEB (Juma 1989). These centres are responsible for the conservation of microbial genetic resources, training activities and the development of microbial technologies appropriate to Third World needs.

The urgency of creating effective transfer mechanisms, including training programmes for scientific and technical staff, is underlined by the limited capability of Third World countries to develop appropriate biotechnologies independently. Thus Busch and Lacy suggest that 'Only a few Third World nations have the critical mass of scientists necessary to engage in genetic engineering . . . None will be able to mount the broad research campaign needed even to stay abreast of First World countries' (Busch and Lacy 1986: 20). Perhaps not surprisingly, newly-industrializing countries, such as Brazil, India and South Korea, are among the few to implement national biotechnology programmes. Inequalities in scientific and technological capability therefore appear inevitable, both between North and South and within the Third World. Without this capability, which needs to build on conventional plant breeding skills, and with inadequate transfer mechanisms, many Third

World countries will not even have the option to develop agri-biotechnologies appropriate to marginal agro-ecosystems and accessible to small farmers. A concerted effort by the international community will be needed to realize this promise.

In the absence of such initiatives, there is significant danger that the deployment of agri-biotechnologies, heavily focused on priorities set by the developed countries, will reinforce trajectories of change in Third World agriculture associated with the Green Revolution and modernization strategies.[23] That is, the trend will be strengthened towards more capital-intensive farming systems, characterized by the diffusion of hybrid varieties, monocultural production and greater genetic uniformity as diversified, self-provisioning peasant agriculture is further undermined. Genetic uniformity also will be accentuated if a limited number of genes controlling agronomically desirable traits (yield, plant architecture, stress tolerance, etc.) are bred into commercial varieties of leading food crops. As the FAO warns, 'there will be a tendency for the more commercially valuable patented genes and gene complexes to accumulate in new varieties' (FAO 1989a: 7).

Furthermore, hybrid varieties hold particular attractions for multinational seed-biotechnology companies competing in Third World seed markets for commercially important crops, such as wheat and rice. Thus Buttel notes that 'the reproductive instability of these hybrids will provide "natural" protection . . . for the proprietary molecules developed through genetic engineering' (Buttel 1986b: 336). Without international agreement on intellectual property systems, genes for specific agronomic traits can be incorporated into hybrids 'without fear that these genes can be duplicated by farmers or competitor firms' (1986b: 336.)[24] Widespread reliance on genetically engineered elite breeding lines for the main food crops would pose a renewed threat to genetic diversity. Without appropriate biotechnologies to improve the sustainability and economic viability of traditional farming systems, competitive pressures to adopt elite cultivars will increase, accelerating genetic erosion. Small farmers, in short, will abandon the role they have performed from time immemorial of conserving germplasm in centres of genetic diversity. 'If this happens, humanity in general will in the long term be the final loser' (FAO 1989a: 9).

CONCLUSION

Agri-biotechnology innovations, whether deployed to enhance sustainable agriculture or promote corporate commercial interests, imply the release into the environment of live genetically engineered organisms (GEOs). Release may involve live bacteria, viruses, plants or transgenic animals, and a number of authorized and unauthorized releases have occurred already, mainly in the United States and the United Kingdom. The direct links with the agri-food system are revealed by field tests to evaluate genetically engineered pest resistance in potatoes and tomatoes, the manipulation of soil bacteria (*rhizobia*) to improve plant nitrogen supply, the spraying of 'ice-minus' bacteria on potato and strawberry plants to lessen frost damage, the introduction of plants resistant to the Monsanto herbicide, glyphosphate or 'Roundup', and tomatoes engineered not to go soft as they ripen in order to extend their shelf-life. These developments, launching the agri-food system on this new course where nature can be 'improved' by design and foodstuffs manipulated, have evoked a variety of responses. One debate concerns the risks associated with the deliberate release of GEOs and the type of regulatory structures needed to avoid undesirable consequences for the environment and consumers. Another current of opinion opposes recombinant DNA (rDNA) research as morally and ethically indefensible interference in the natural order of life and the process of evolution. This polarization reflects fundamental philosophical differences between the Baconian positivist/rationalist tradition and more spiritual, holistic approaches, which challenge the means *and* ends of genetic engineering. These question the morality of genetic manipulation and the hazards of embarking on this final step in humankind's control of the genetic secrets of all living organisms.

Although most industrial countries have established safety guidelines for laboratory rDNA research,[25] progress in creating regulatory structures to control environmental release of GEOs is uneven and incomplete.[26] Certainly statutory structures are urgently needed in view of unauthorized releases in the United States, and the field testing in 1986 of a live rDNA rabies vaccine on cattle in Argentina by the Wistar Institute of Philadelphia without the knowledge of the local authorities. The central problem in devising regulatory procedures concerns the potential hazards which may arise from the movement of recombinant genes through the environment due to genetic transfers between micro-organisms and between plants via cross-pollination. The 'scientific community' is divided on the magnitude of the environmental risks such gene transfers present and the rigour with which they should be assessed.

These divisions emerged strongly at the 1988 Cardiff conference on the release of genetically engineered micro-organisms into the environment (REGEM). Ecologists in particular stressed the serious inadequacy of current understanding of microbial ecology and the behaviour of ecological systems as a basis for environmental risk assessment (ENDS 1988). Such gaps rule out comprehensive surveillance of the movement of novel organisms in the environment, whether their genetic effects are benign or harmful. After the debates of the 1970s on the laboratory containment of GEOs, Professor Hans Kornberg, chair of the UK Advisory Committee on Genetic Manipulation, is reported as saying that with planned environmental release 'we are once again facing an unknown and as yet unknowable situation' (ENDS 1988: 16).

Most scientists acknowledge the weak foundations of ecological risk assessment and recognize that genetic transfer will probably occur as greater numbers of novel organisms are introduced into the environment. Yet not all microbiologists equate this with the need for stringent safeguards. Some self-styled 'progressives' argue that the risks are acceptably low and, in any case, are far outweighed by the benefits (ENDS 1988). However, regulatory authorities in the United Kingdom appear to have opted for a cautious proactive approach, and the Royal Commission on Environmental Pollution recommended in 1989 that 'release licences' be granted initially only on a case-by-case basis.[27] Nevertheless, this falls far short of the moratorium on all releases demanded by environmental groups, such as Greenpeace, until there is better scientific understanding of the environmental consequences. In conjunction with belated efforts to improve knowledge of microbial ecology,[28] ways of restricting genetic transfer also are being explored, including the insertion of biological safety mechanisms and 'suicide genes' into manipulated organisms (ENDS 1988).

Regulatory authorities in Western Europe and the USA so far have resisted pressures from public interest groups to go beyond 'technical' considerations of environmental safety and incorporate social and ethical criteria when evaluating applications for planned release. One proposal is to assess each product under review in terms of social or market 'need'. This criterion gained widespread support in Europe during the controversy over the licensing of bovine somatotropin (BST), a synthetic analogue of a natural hormone which is injected into cows' udders to increase milk secretion.[29] In similar vein, the UK Genetics Forum, a public interest group created in 1989, has called for a ban on the development of herbicide resistant plants. Such groups are demanding less secretive and more democratic regulatory structures, including in the United Kingdom the formation of an independent national committee equivalent to the Warnock Committee, which would examine the ethical and social implications of biotechnologies and genetic engineering of plants and animals.[30]

A full discussion of these issues would take us far beyond the scope of this volume. Nevertheless, there can no longer be any doubt that the restructuring of the agri-food system by biotechnologies raises questions of the deepest possible significance for the farmed and wild environment. Nature is being refashioned for commercial gain and human purposes in ways which transcend the capacity of nature itself to achieve. The relationship between the agri-food system and the environment thus takes on dimensions which have yet to be revealed and whose irreversible consequences are unknown and unforeseeable.

NOTES

1 The potential scope of industrial applications of molecular biology through genetic engineering and biochemical engineering goes well beyond the food system (Bull *et al.* 1982; OTA 1981). An incomplete list would include the energy industries, chemicals, pharmaceuticals, waste treatment and pollution control.
2 This discussion concentrates on plant biotechnologies, although commercial appplications of animal biotechnologies are further advanced, including vaccines, in vitro fertilizatation and embryo transfer, growth hormones and feed additives. The genetic engineering of animals also

has more profound ethical implications in view of arguments that mammals possess moral rights.

3 Plant genetic engineering so far has concentrated heavily on single gene traits but many desirable agronomic characteristics are determined by a number of genes. This focus reflects serious gaps in scientific understanding of how multigenic traits are expressed.

4 These factors evidently were decisive considerations in the agreed take-over of the California-based Genentech, the largest and most successful biotechnology firm, by Hoffman-La Roche in February 1990.

5 This issue of access was raised in an acute form in November 1989, when James Watson, the leader of the American project to map the human genome, surprised the scientific world by announcing that the United States would deny research information to countries not contributing funds to the collaborative research programme. This restriction flies in the face of scientific tradition and is a further indication of the commercial ethos taking over molecular genetics research.

6 These soil bacteria have a symbiotic relation with some plants, forming nodules on their roots and transforming atmospheric nitrogen into nitrates, which feed the plant.

7 These cuts followed an internal Cabinet review, the so-called 'Barnes Review' in February 1988, which determined the withdrawal of £30 million from agricultural R&D considered to be 'near-market', a reduction of roughly 30 per cent in public research funds (Read 1989; Webster 1989).

8 Gaps in the national science base will also appear if the withdrawal of public funds from 'near market' research related to commercial products is not met by a compensatory expansion of industrial R&D expenditure.

9 Noting that the Pasteur patent granted in the United States in 1873 included a claim on a cultured micro-organism, Juma (1989: 161–2) observes that 'The patenting of life forms has co-evolved with other forms of intellectual property protection and what is happening now is a convergence of developments that have taken place differentially in other fields.'

10 These tendencies have attracted increasing opposition to genetic engineering on ethical, environmental and social grounds. The law suits brought against biotechnological firms in the United States by Jeremy Rifkin and the Foundation on Economic Trends are widely known. In Europe, the March 1989 decision of the European Patent Office (EPO) to award the American firm, Lubrizol, the first European patent on a genetically engineered plant has been challenged by a coalition of public interest organizations. In this important test case, these groups contend that the European Patent Convention of 1977 is inapplicable to living organisms. Also in 1989, the EPO rejected a patent claim by Harvard University on a transgenic mouse (Kingman 1989).

11 Kloppenburg (1988a) notes that seed firms have acquired relevant experience in enforcing the PVPA of 1970.

12 For example, in response to the campaign waged by the US Pharmaceutical Manufacturers Association for commercial sanctions, Chile introduced legislation in January 1990 to recognize pharmaceutical patents.

13 On 'genetic imperialism' and the creation of 'botanical empires', see Mooney (1979; 1983), Kloppenburg (1988b) and Juma (1989).

14 In addition to debates about the asymmetries and alleged inequities of this system, the technical problems of maintaining genetic diversity through *ex situ* collections are now widely acknowledged. This has increased support for the creation of conservation areas in centres of genetic diversity to ensure the survival of wild relatives of commercial crops and the primitive landraces cultivated in peasant agriculture.

15 Unlike the utility patent protection conferred by the 1985 *Hibberd* decision, plant breeders' rights extend only to new crop varieties, recognizing the breeders' monopoly for a limited period, and do not cover the genetic material they contain nor the breeders' techniques *per se* (FAO 1989a).

16 The view that common heritage in plant genetic resources is both a 'quixotic pursuit' and likely to bring fewer benefits than national-property initiatives and compensatory mechanisms is defended by Kloppenburg (1988b).

17 These authors observe that 'The driving force behind recent interest in developing large-scale production methods for plant tissue culture-derived products has been the realization that in various plant species . . . the targeted product may be produced in concentrations severalfold higher than in the intact plant' (Sahai *et al.* 1986: 76).

18 In Western Europe, sugar beet growers successfully lobbied for the introduction in 1977 of an EC regulation restricting output of HFCS and other cereal sweeteners.

19 According to Singh (1989), sugar exports from the Philippines to the United States declined from $624 million to $246 million between 1980 and 1984, and half a million jobs have been lost with the shift to less labour-intensive crops.

20 These include specialist industrial chemicals and ethanol, emphasizing the emerging technological base of a new food–chemicals–energy complex using renewable feedstocks. Examples are the US corn industry discussed here and the Brazilian sugar complex, with its *Proalcool* programme and incipient sucro-chemicals industry.

21 A key, and as yet unresolved, issue for plant improvement work, including international collaboration, is whether or not new patent legislation will incorporate a 'research exemption' to give breeders ready access to genetically modified germplasm.

22 Deo and Swanson argue that the influence of OECD governments on the CGIAR system led to the systematic neglect or de-emphasis of 'alternatives to the Green Revolution, including non-irrigated, labour-intensive methods of cultivation, dry farming practices, biological means of disease control and prevention, etc.' (Deo and Swanson 1987: 28).

23 Although private biotechnology firms are pursuing R&D on 'bioinsecticides', 'biofertilizers', and other products that will lessen dependence on traditional agricultural chemicals, 'the development of crops tolerant of widely-used herbicides (e.g. glyphosphate, triazine) represents the single most common research programme in private firms' (Buttel 1989b: 15).

24 The caveat here is that it may be possible to clone hybrid genotypes using tissue culture and thereby avoid buying new hybrid seed for each planting (FAO 1989a).

25 Earlier controversies on the morality and biological hazards of rDNA work, including the 'Berg letter' of 1974 and the 1975 Asilomar Conference, are discussed by Kloppenburg (1988).

26 In the case of the United Kingdom, see Royal Commission on Environmental Pollution (1989) and Shackley (1989).

27 At the time of writing, US regulatory procedures are administered under existing statutes by established agencies, such as the EPA and FDA, and focus on the product and product category rather than the process involved in its production, as in the United Kingdom. A current test case of this approach concerns the FDA's decision on whether the 'super' non-rotting tomato developed by Calgene is just another tomato or an entirely new plant whose dissemination should be carefully monitored. Appeals by private firms for limited public regulation contrast sharply with the claims of novelty made in patent applications.

28 In the UK, substantial funding for research into the behaviour of GEOs released into the environment has been made eight years after the first such release at Rothamsted Experimental Station (Shackley 1989). The three-year programme PROSAMO (Planned Release of Selected and Modified Organisms), announced in August 1989, is jointly funded by the Department of Trade and Industry, the Agricultural and Food Research Council, and a consortium of eight companies.

29 According to press reports in February 1990, seven major chemical companies – ICI, Ferruzzi, Hoechst, Monsanto, Rhone-Poulenc, Sandoz and Unilever – delivered a joint paper to the European Commission expressing their concern about plans to incorporate a market need criterion in EC approval procedures for biotechnology products. They reportedly observed that it might be difficult to demonstrate for agricultural products at a time of commodity surpluses. The companies also appealed to national and Community sentiment, arguing that such measures would put European competitiveness in biotechnologies at risk.

30 This proposal was made at a meeting organized by the National Consumer Council in October 1989 (*The Independent*, 9 October 1989). Dame Mary Warnock chaired the Committee of Inquiry into Human Fertilization and Embryology set up in July 1982.

6

THE FOOD SYSTEM AND THE ENVIRONMENT

Refashioning the genetic structure of crops and animals, important as it is, represents only one face of the technological revolution which has marked the appearance of the modern food
system. Just as nature has been redesigned from within, through
animal genetics and plant biotechnology, it has been transformed
from without, by leading, in turn, to changes in the physical
environment, or countryside. These changes in the 'face of the
land', themselves stimulated by technological change and accompanying shifts in the labour process (as we saw in Chapters 1 and
2) have contributed to new social conflicts and political demands
around what has come to be seen as the Green agenda. The
development of the food system implies a major change in the
'natural' environment, as well as in the genetic composition of
the species. The scope of this change can hardly be overestimated.

The environmental effects of modern agriculture point to
contradictions within the entire model on which the modern food
system is based. Modern agriculture developed out of the partial
separation of 'farming' activities from the natural resource base
on which these activities were traditionally dependent. As we saw
in Chapters 3 and 5 this led to technological innovation, and
accompanying policy incentives, which have presented problems
for the planned evolution of the food system. In some sectors,
such as the production of chemical fertilizers, plant breeding
and pesticides, innovation developed 'autonomously', according
to a scientific and industrial logic only loosely related to the
cycle of agricultural production. The reintroduction of new
industrial components into agriculture, and the food system
more generally, posed problems in two ways: some processes
were taken out of 'farm' production (appropriation) and others

were substituted, especially in food processing activities (substitution).

These transformative technological changes were associated with major shifts in the labour process, not only on the farm but in the household and in the food industry. These shifts have been the focus of attention in Chapter 1, and again in Chapter 5. However, an area of equal concern has emerged during the last decade or so, which promises to pose even more serious problems for the evolution of the food system: the environment. Recent decades have witnessed a series of policy interventions designed to reduce, or reverse, the unforeseen or negative effects of the food system on the environment, largely in the industrialized countries. No area of concern demonstrates the difficulty of managing the contradictions of the food system as clearly as the environment.

Conflicts over the rural environment, and over food quality, have come to represent the clearest indication yet of the problems which the modern food system poses for sustainable development, and for the model of economic growth based on intensive accumulation which, as we have seen, spawned the modern food system itself.

This chapter examines the impact of the modern food system on the environment, in both the developed and developing countries, and the social responses which these environmental changes have evoked in recent years. The analysis begins by considering the changes within the food system which have undermined long-term sustainability, changes which were associated with the transference of economic activities between the farm and other components of the wider system. In the next section the major agricultural policy instruments which have been brought to bear on the rural environment in the industrialized countries are examined, notably the effects of price-support policies and economic incentives in the introduction of more capital intensive, 'high'-technology agriculture. The following section considers the effects on the environment of these policy mechanisms, including the loss of genetic resources, the pollution of water and soils, the threats to sustainable utilization posed by land degradation and the effects of agricultural change on amenity and recreation in rural areas. These environmental effects of the modern food system are considered as first-, second- and third-order problems, each of which requires major adjustments, at the international as well as the national level. Finally, the chapter examines the role of management interventions and environmental movements in seeking to modify or reverse these processes. A conceptual framework is outlined, which explores different forms of 'resistance' to the development of the modern food system, and concentrates attention on the epistemological and political aspects of these responses.

THE FOOD SYSTEM LEAVES THE FARM

The description of farming in Hooten Pagnell in Chapter 2 contains most of the elements of what was still, in the 1930s, looked upon as a 'natural' farming system, heavily dependent on the cyclical and seasonal quality of rural life. Animals were left to graze in the fields, or fed on fodder crops produced on the farm, bought inputs such as fertilizers and pesticides played a negligible part in the farm accounts, a sizeable part of the food for home consumption was still provided by the farm, marketing was still in its infancy, and the 'farm family' (largely the women) processed and prepared food in the home. Each of these elements, as we have seen, has been removed (partially or fully) from the farm economy and relocated, usually in an industrial form, elsewhere in the food system.

In contrast to this picture, modern agriculture, as we have seen, is dependent on very high levels of protein and energy transfer, linking the production process on the farm with sectors in which value is added to farm production. These linkages, both upstream and downstream, have broken into the cycle of renewal, failing to return to the resource base what has been removed from it, a principle which marks sustainable agricultural systems. Thus, in modern farming, animal waste disposal has become a 'problem' rather than an essential part of a natural recycling process. In the same way, advanced systems of crop and animal husbandry, using every protective device from pesticides to antibiotics, are made more necessary by the conditions under which animals are kept, and the genetic make-up of delicately engineered plant varieties. Given the commitment to high yields, consistent quality, and the major investments in research and development which underpin these technologies, once they are established there is apparently no cost-effective alternative to capital-intensive, energy-wasteful production regimes.

Nevertheless, in the process of establishing these regimes the 'knot' which tied agriculture to the environment has been undone, and the costs of modern agriculture are counted elsewhere. In the United States for example, it has been estimated that the cost of soil erosion off the farm was US$6.1 billion in the mid-1980s. Similarly, in the United Kingdom, £200 million in capital costs and an annual expenditure of £10 million will be required, over a twenty-year period, to bring all drinking water up to the European Community's standard for nitrate concentrations (OECD 1989). In the developing countries the so-called 'externalities' are often higher still. Repetto et al. (1989) have argued that although gross domestic product (GDP) in Indonesia grew at an annual rate of 7.1 per cent between 1970 and 1984, if this was corrected to take account of the depletion of timber, petroleum and top soil, the rate of economic growth falls to only 4 per cent, and for four of the fourteen years was actually negative. Similarly, by calculating the effects of soil erosion on agricultural growth over ten years, Bishop and Allen (1989) have shown that in Mali soil erosion accounts for up to 16 per cent of agricultural GDP. Costs of this sort are not included in the national accounts, and hence calculations about the benefits of agricultural growth rarely reflect the real losses sustained over long periods.

There are several important respects in which attention has shifted to the problems the modern food system presents to the environment. First, the increased dependence on differently located components of the agricultural production and food processing system has served to transfer environmental costs from one part of the system to another. For example, as we have argued elsewhere, the use of soya meal and manioc for intensive cattle production in Europe, has transferred some of the costs of maintaining feedstock systems to other countries, such as Brazil and Thailand, which can ill afford it. Second, increasingly intensive production has led to problems of waste disposal, pollution and disease control, all of them problems which have accompanied intensification. These problems frequently assume major importance in two directions: for the farm they represent increased costs which need to be incorporated in the farm accounts, and the market viability of the enterprise; and for the local environment they represent major threats to the existing ecological systems, species diversity and habitats. With increasing demands for financial security, but tighter economic margins, expenditure on what are termed 'intermediate goods' now represents nearly half the value of agricultural output of some industrialized countries such as France, as Table 6.1 makes clear.

Table 6.1 Intermediate consumption in French agriculture (billion francs)

	1970	1975	1981
Animal feed	6.8	14.9	33.0
Fertilizers	4.2	9.0	19.1
Fuel and repairs	3.2	6.8	15.4
Crop protection	1.2	2.9	9.6
Building maintenance	0.8	1.4	3.2
Veterinary costs	0.9	1.9	4.5
Other goods	2.5	4.8	9.5
Other services	1.4	2.4	5.0
Total intermediate consumption as proportion of output	30.0%	36.0%	46.0%

Source: European Parliament Working Document A 2-207/85 (1985–6), Report on behalf of the Committee on the Environment, Public Health and Consumer Protection.

This table, which expresses the costs of intensification both in current money terms, and as a percentage of total production costs, clearly points to the increasing financial burden of maintaining intensification, which is borne by the farmer in the first instance, but indirectly by the consumer and taxpayer and, at another remove, by agricultural competitors in less protected markets. As we shall see later in the chapter, the environmental costs of this pattern of financial and resource management on the farm are paid in a number of ways, not all of which are easily 'accountable' in monetary terms.

The third aspect of the removal of 'natural' processes from the farm, and the relocation of these processes in other sectors of the food system, has been to transform, in some cases fundamentally, the pattern of land use in rural areas. The modern food system has brought unprecedented interference with traditional land uses and the rural landscape, especially through activities like drainage, land levelling and deforestation.

There has also been a reduction in soil quality, arising from the poor management of soil resources (European Community 1985–6: 13). Current problems, as we shall see, include precipitation of acid rain on land, the discharge of toxic wastes and the uncontrolled spread of sewage on land. As a result of these and other factors, the usable agricultural area within the European Community declined by almost 8 per cent between 1962 and 1984, from almost 100 million hectares to under 92 million hectares (1985–6: 15). The combination of several different processes, whose cumulative effect is to undermine the quantity of undegraded land available for both agricultural and non-agricultural uses (including conservation) has led to a rural resource crisis in Europe of serious dimensions. As the Food and Agricultural Organization (FAO) of the United Nations has put it, in rather august language:

> A complex web of economic policies provides, at considerable cost to the environment, many of the incentives and institutional mechanisms needed to support the high level and intensity of agricultural production. This situation is compounded by over-production of some commodities in the agricultural sector and rapidly rising economic costs from policies intended to encourage production and raise income.
>
> (FAO 1988: 1)

It is to this 'complex web of economic policies' that we need to turn next.

THE PRODUCTIVIST MODEL:
POLICY INSTRUMENTS AND MECHANISMS

In a document by the European Parliament itself, we are reminded of the lack of attention to long-term sustainability in the development of the European Community's agricultural policy (European Parliament 1985–6). Emphasizing the five fundamental objectives of the Common Agricultural Policy (CAP) of the European Community (article 39 of the Treaty of Rome) – to increase agricultural productivity, improve farm living standards, stabilize markets and ensure a secure supply of foodstuffs at a reasonable price – the document continues by stating that 'the pursuit of these objectives clearly betokens a productivist and materialist ideology which takes no account of environmental concerns' (1985–6: 19). Community preference follows logically from the establishment of a single agricultural market, but it is also clear that establishing this market is expensive in terms of export subsidies 'and that it does not necessarily contribute to the rational location of different types of production' (1985–6: 19). Similarly, within the European Agricultural Guidance and Guarantee Fund (EAGGF) of the European Community, 'guidance' to farmers (which might assist with the painful process of restructuring for farm households) only amounts to one twentieth of the total expenditure: the bulk of expenditure under this fund takes the form of subsidies to agricultural producers.

The scale of agricultural subsidies, as we shall see, makes any serious attempt at environmental protection in European agriculture extremely difficult, since farmers who are normally subsidized for doing what is harmful to the environment also expect to be subsidized for not doing what is environmentally harmful. The cost of the CAP is, as Simpson (1989) remarks 'legendary'. Only part of the total cost of the CAP is attributable to the purchase, storage and export subsidies for farm production: a 'mere' £18 billion per annum, equivalent to £5.91 a week for a family of four in the Community. In addition, the same family has to pay almost twice this amount weekly towards the cost of fixing EC prices above prevailing world market levels. The total cost is £42 billion per annum. These figures provide an indication of 'a policy which sacrifices long-term interests to immediate concerns, and one scarcely needs reminding that most environmental issues arise from a concern for the long-term future' (European Parliament 1985–6: 19).

It is worth paying attention to the policy framework for agriculture within the European Community, because in intention, although not always in its choice of mechanisms, it mirrors the approach of other industrialized countries, such as the United States and Japan. The policy framework for European agriculture, as we have seen, was established with a view to protecting rural incomes through production subsidies, and ensuring food sufficiency within member countries. One consequence of these policies is that the benefits and costs of agricultural policy are shared very unequally between producers and consumers. For example, if we compare France with the United Kingdom, we can appreciate that because of the larger size of the French agricultural sector, the absolute value of protection is higher than in the United Kingdom. However, if we take individual farmers the reverse is true: British farmers do rather better under the CAP because they work, on average, larger farms. In France in 1980 the average farm size was 25.4 hectares compared with 68.7 hectares in the United Kingdom. As Cheshire (1985) makes clear, 'large farmers, because they produce more, effectively receive more support than poorer, smaller farmers' (Cheshire 1985: 11). The aggregate effect of these distributional factors is highly unequal: three-quarters of EC support to farmers goes to the richest quarter of farmers (Simpson 1989: 15).

The level of expenditure under the CAP far outstripped most other budgetary items within the control of the Community. Taxpayers and consumers effectively supported 40 per cent of the value of farmers' produce during the years 1983–7, when agricultural spending in real terms increased by 4 per cent a year. In the United Kingdom, in 1984, more was spent on farmers and the food industry (£6.9 billion) than was spent on unemployment benefit and supplementary benefit combined (£5 billion). Senior (1989) also points to the fact that only a dozen people work for the European Commission on consumer affairs, while 800 work in DG.VI, the directorate responsible for agriculture. And within the European Community it is DG.VI that makes agricultural policy, in consultation with the agricultural producers' group COPA, the Council of Agricultural Ministers and the Agricultural Committee of the European Parliament.

The effect of close collaboration between farming interests, and the virtual exclusion of the consumers' interest from food policy discussions, is that policy not only favours large farmers over smaller ones, it also favours wealthier families over poorer ones. In 1988 in the United Kingdom the poorest 25 per cent of households spent 29 per cent of their disposable income on food. The richest 25 per cent of households spent 14 per cent on food. The relatively 'rich' West Germans spent 14 per cent on food, while the relatively 'poor' Greeks spent 36 per cent. Simpson observes that 'if poor people were paying to help other poor people and keep them on the land that may be some consolation. But they are not' (Simpson 1989: 15). Very little influence is wielded within European Community circles by the consumer interest, and environmental considerations still do not enter into the design of agricultural policy, as one might expect. As we have seen, this lack of political influence is in marked contrast with the position of the agricultural lobby. In explaining the ideological dominance of this lobby, we must take account of the level of farm support in the past. The level of farm support has been fundamental to the shift in agricultural practices, most of which, as we shall see, are prejudicial to the environment.

It is not difficult to demonstrate that the mechanisms, and the level, of farm support in the European Community is highly inegalitarian in its effects. Can it be demonstrated that these policy mechanisms are ultimately responsible for the environmental problems that afflict the countries of Western Europe?

The cumulative effects of high levels of farm support on the rural environment may be complex and subtle, but they are also increasingly well documented. As we saw in Chapter 3 the agricultural technology/policy model encouraged the rapid adoption of new technology in the United States and Western Europe. The high level of farm support, expressed through high farm prices, has encouraged more investment in research and development, and the acquisition of more capital equipment by farmers. Price support has also stimulated intensification through its effect on land values. Throughout the period between the mid-1950s and the late 1970s agricultural land values increased in the United Kingdom, in fact they tripled in real terms. After 1979, when the rate of increase of CAP price began to fall, and particularly after 1984 when milk quotas were introduced, land prices began to decline, for the first time in decades. The intensification of agriculture was given a considerable stimulus by the capitalization of land as an asset, and the need to maximize returns on this asset.

It is also clear, from evidence such as that of Traill (1983) that high support prices for farmers induce more capital investment, and the substitution of capital equipment, where possible, for labour. In West Germany, in 1989, there were 8.5 milking machines for every 100 cows; 2.8 combine harvesters and 12.4 tractors for every 100 hectares of land. This level of mechanization was higher than in the United Kingdom because the farm structure was different: under a regime of high support prices, and accompanying subsidized credit and guaranteed markets, there is every inducement for even the smallest farmers to mechanize production.

Second, although the legal and financial inducements to farm amalgamation are few within the European Community, the economic gains from larger units, from 'shedding' labour and acquiring ever more capital equipment, are considerable. The pressures on the production margins of small farmers, even those with milking machines and tractors, are further accentuated by the concentration of farm ownership. One effect of agricultural support and the technological treadmill is, therefore, greater specialization. It is in the automatic consequence of inducing investment, and losing employment, in farming.

FIRST-ORDER PROBLEMS WITH THE PRODUCTIVIST MODEL:AGRICULTURAL INTENSIFICATION

The productivist model, which lies at the heart of the modern food system, has both direct and indirect environmental effects–environmental problems in the immediate locality, and others at one remove. The environmental problems associated with agricultural intensification are, in the first instance, experienced as a direct effect of the implementation of the agricultural technology/policy model. As we shall see, these first-order problems hold for both developed industrial societies and for areas of modern agriculture within developing countries.

Developed countries

The effects of agricultural intensification can be illustrated by taking as an example the removal of hedgerows (Cheshire 1985: 14). Similar accounts could be given of the effects of drainage (Baldock 1984), or the application of chemicals or intensive livestock production (Long and Rose in Goldsmith and Hildyard 1986).

Most factors of production are in relatively elastic supply, but land is the exception. The supply of land can normally only be increased through 'improvement' or intensification. As agriculture feels the effect of greater capital investment, prices increase and these become capitalized in land. One effect of this process is to increase the opportunity cost of unproductive land, such as hedges and field margins. The farmer will view the conservation of hedgerows as an opportunity foregone – he has forfeited the opportunity to increase his income by producing more food. This tendency to regard conservation activities as a cost to current income is exaggerated because, under the system of grants to defray capital costs within the European Community, farmers are induced to specialize, substituting capital for labour in many of their operations. As Cheshire argues, 'while labourers could milk cows, tend the sheep and with a simple binder, harvest the corn, milking parlours can only milk cows, and combine harvesters can only harvest cereals' (Cheshire 1985: 15). By increasing output the farmer was induced to replace organic with chemical fertilizers, and to use chemical sprays for pest control. Both sprays and fertilizers depend on subsidized machinery.

Under specialized production hedgerows were unnecessary, since they served no purpose as livestock barriers, and animals had been brought in from the fields. The use of larger machinery also meant that fields needed to be larger, persuading farmers to recoup their increased capital costs by removing unproductive hedges and ploughing over footpaths, which interfered with the movement of their mechanical equipment. The environmental problems that have accompanied intensification are the 'effect of a system which systematically establishes financial inducements to erode the countryside, offers no rewards to offset market failures and increases the penalties imposed . . . on farmers who may want to farm in a way which enhances and enriches the rural environment' (Cheshire 1985: 15). The example of hedgerow removal is just one of many similar examples of foregone production benefits being converted into environmental losses.

Among the most important effects of intensification is the use of nitrogenous and other chemical fertilizers. Within the European Community the use of chemical fertilizers has assumed increasing importance. According to the FAO, consumption of nitrogenous fertilizers in the Netherlands is 558 kg per hectare, Switzerland 436 kg, West Germany 427 kg and the United Kingdom 356 kg (FAO 1988: 5). There is evidence that these levels of use leave significant residues in the soil. A report on Paris Basin agriculture shows a nitrogen surplus of 240 kg per hectare over a seven-year period. Eutrophication of lakes, rivers and coastlines caused by excessive nutrient loading from nitrate and phosphate pollution is also widespread. The FAO notes that recent acute deterioration of marine ecosystems in the Kattegat and the Adriatic Sea, for example, has had a significant impact on fisheries and tourism in these areas.

The intensive production of animals in feedlots has had similarly important environmental effects. It is estimated that each year British farmers dispose of 200 million tons of animal manure through the river system. The disposal of slurry is now a major environmental problem. A survey by the Water Authorities Association in July 1986 revealed that incidents of silage effluent pollution were almost double those of two years earlier. In 1985 Anglia Water reported that half of all fish deaths leading to prosecutions, and 60 per cent of its pollution prosecutions, were attributable to slurry disposal. The excessive use of animal slurry can lead to concentrations of up to 10 kg of nitrogen per hectare. The mineralization of the soil from increased large-scale nitrate leaching can pass into both surface and groundwater. Already, in the Netherlands, the nitrate content of shallow sources of fresh water is too high to make it usable (European Community 1985–6: 32). The average nitrate content of aquifers in the Paris Basin has doubled from 20 to 40 mg per litre during the last twenty years, while in some areas of France, such as Beauce and Sologne, the nitrate content of groundwater exceeds 100 mg per litre. Studies of more than 6,000 wells in North Rhine–Westphalia showed that two-thirds had nitrate contents of 50 mg per litre or above, the maximum permitted under EC Directive (80/778, 15 July 1980). A third of the wells exceeded the very lenient limit of 90 mg per litre permitted in West Germany. The cost of reducing nitrate concentration in groundwater to within acceptable limits in West Germany has been calculated as 2000 DM per hectare (Vogtmann 1985: 5). Facts such as these have led to alarm within EC policy circles. European Commission guidelines state that: 'indefinite exemption from the principle that the ''polluter pays'' can hardly be given to agricultural holdings, least of all those which, in terms of the way they are structured and operated, are to all intents and purposes industrial undertakings' (European Commission 1986: 20).

The conversion of forests, pastures and wetlands to other uses has led to environmental losses in other directions, notably the loss of wildlife habitats, hydrological changes and increased risk of erosion. Land reclamation practices have brought about radical changes in the environment during the last decades and, although the rate of land conversion is decreasing within the European Community, it is still a contributory factor in further environmental deterioration. In the United Kingdom, for example, a dramatic loss of valuable biotopes and wildlife habitats has occurred since 1945. Almost all unimproved natural grasslands, including herb-rich hay meadows, have disappeared, as have 40 per cent of lowland acidic heaths, 30–50 per cent of lowland woods and 30 per cent of unimproved upland grasslands, heaths and blanket bogs. As the FAO report to the United Nations Economic Commission for Europe expressed it, 'the combined effects of modification of traditional agricultural landscapes, increased use of pesticides and intensification of agriculture have reduced wild species abundance and diversity in most agricultural areas [in Europe]' (FAO 1988: 4).

There are many interrelated reasons for loss of soil productivity which result from agricultural intensification and heavy reliance on external inputs. These effects are not confined to developed countries, like those of Western Europe. The more or less proximate effects of agricultural intensification are also felt in areas of the Third World where modern agriculture has become established, often with fewer controls on adverse environmental effects than exist in the developed countries. These areas, which we can term 'enhancement zones', perform a critical role in food production in the South, and particularly the production of food staples such as rice, which are grown under highly intensive irrigated conditions. The changes discussed below, through which modern intensive agriculture has served to transform the environment of developing countries, suggest that the modern food system does not need to be fully elaborated within developing countries, from both a consumption and production standpoint, for new agricultural technologies to carry important social and environmental implications there.

Developing countries

In the developed countries the consequences of the modern

food system have been challenged on the grounds that financially underwriting modern agriculture has proved extremely expensive: the costs of agricultural support are paid by the taxpayers and consumers within the European Community, whose interests are now beginning to be championed by political leaders as well as lobbies. The other major objection to agricultural intensification has come from the environmental groups opposed to high-input farming, or interested in preserving the amenity and recreational aspects of the countryside.

Government support for agriculture in the developed countries has been financed from domestic budgets, and the financial costs have been passed on to the population of these countries. In the developing countries the circumstances are rather different: new developments in technology and the implementation of agricultural development policies, have been financed through external loans and credits. For example, the International Agricultural Research Centres, which form part of the CGIAR system discussed in the last chapter, are largely supported by individual governments in the developed world, development agencies such as the World Bank and by large foundations such as Rockefeller and Ford. To the extent that these support costs are visible, and many of them are not, it is difficult for the people of Third World countries to mount an effective opposition to the way funds are disbursed on their behalf. Similarly, people throughout the developing world are often very aware of the environmental costs of modern agriculture in areas of intensive production and of the environmental costs in 'low resource areas' which have been increasingly marginalized from the development process. However, as we shall see, in most developing countries it is much more difficult to establish environmental priorities in the face of the 'gains' expected from economic growth, including gains in agricultural production. Yet the heavy cost of environmental losses should not be underestimated, as we have seen. Developing countries are effectively caught in a 'scissors movement'; the costs of conserving natural resources are borne by them, while they also bear the cost of revamping their economic policies under the exigencies of structural adjustment, and other packages emanating from international financial institutions.

As we saw in Chapter 4, in many parts of the Third World, especially Asia, the Green Revolution has often been depicted as a success of 'science' and 'technology' in overcoming the obstacles to increased agricultural production. In fact, as a number of writers have indicated (Biggs 1988) the most consistent factor in the production successes attributed to the Green Revolution has been public sector investment, often reinforced by externally-secured finance, and including a series of support measures affecting prices, marketing arrangements and crop procurement, the transfer of technology and input manufacture. The effects of the modern food system have been felt, if not directly then indirectly, in the way developing countries have sought to subsidize the adoption of increasingly intensive agricultural packages. The 'successes', in production terms, of the Green Revolution in most areas where wheat and rice are staple crops can be attributed in large measure to the transfer of the agricultural technology/policy model first developed, in its modern form, in the United States in the 1950s and 1960s. In the developing countries the International Agricultural Research Centres, in particular, performed the kind of role discussed in Chapter 3, under which research institutions in the United States developed and transferred new technologies from the laboratory to the land.

The adoption of these technologies, in the developing countries as well as in the North, had more to do with the incentive structure that surrounded both their use on farms and the research and development exercise. One important aspect of the expansion of the modern food system is that components of this system, if rarely the system in its entirety, have served to transform the context of agriculture in so many Third World countries, despite structural constraints to the adoption of the fully elaborated industrial food system. Essentially, the adoption of 'high' capital-intensive technology under conditions in which most of the rural population live in material poverty, has in many cases served to exacerbate inequalities within regions. Under these circumstances, ecological problems, especially increasing fragility of the ecological system, are closely related to a reduction in livelihood choices for poor people, and their efforts to re-establish a viable livelihood are frequently costly in environmental terms, whether in resource-poor areas or in areas of capital-intensive agriculture (Redclift 1990 discusses this in more depth).

215

Recent papers from the World Bank have suggested some useful policy options, which would enlarge the management interventions available to government. One example is the removal of pesticide subsidies, which are costly in economic terms, for most developing countries, and undermine the stability of agriculture. The removal of these subsidies, or the introduction of a pesticide tax, would reduce pesticide consumption, and give farmers an increased incentive to adopt more sustainable, biological alternatives (such as those indicated under the general rubric of Integrated Pest Management). In the case of Indonesia, it has already been demonstrated that reducing pesticide subsidies brought fiscal savings of US$150 million annually over a period of four years (Brady 1989). As World Bank economists have shown, the reduction of these subsidies contributed to more pest-resistant rice varieties becoming available, and Integrated Pest Management practices became more widespread (Schramm and Warford 1989: 16).

SECOND-ORDER PROBLEMS WITH THE PRODUCTIVIST MODEL: MARGINAL LANDS

The secondary, or indirect, effects of the food system on the environment in both developed and developing countries are more difficult to delineate. The process through which the modern food system becomes established at the farm level – concentration of holdings, specialization and intensification of production – carries implications for areas other than those of modern agriculture. The cost of achieving gains in productivity in one locality is often passed on to others; through environmental costs or 'externalities'. As we have seen, intensification needs to be understood not simply in terms of the technology employed but also in terms of a 'package', comprising guaranteed prices, subsidized chemical and mechanical inputs and government procurement after the harvest.

In some developed countries, such as those of the European Community, less advantaged regions receive subsidies from central government precisely because they are adversely affected in social terms – they are less competitive in terms of their economic performance. In developing countries, however, areas where natural resources are poor, especially semi-arid regions, are likely to fall completely outside the ambit of public policy. This neglect serves to accentuate the differences, the unevenness of the development process. It is more difficult for small farmers, for example, to gain government support for their farming system; they are left to manage an inadequate technology, even if this technology is environmentally sustainable, ill-adapted for competition with the technology employed in more favoured regions. Many households, in marginal areas of Europe, Australia and North America, such as the Appalachian Mountains, as well as the Third World, have to supplement the income they obtain from their land by other forms of livelihood. They become part-time farmers, working in the city or in tourism, or migrating on a seasonal basis to other regions of the country. Living in areas that are already poor in terms of productive resources, but frequently rich in terms of natural species or aesthetic quality, they are powerless to do anything in the face of the differentiating effects of modern intensive agriculture.

This chapter is concerned with both the immediate and the more indirect environmental implications of the modern food system. It needs to be emphasized that a series of 'third-order problems' also exist which we are only beginning to delineate: that is, problems such as those arising from the introduction of genetically engineered species into the environment. In the previous chapter, the discussion of biotechnology and genetic engineering touched on these issues, but it is too early to elaborate at length on the social implications. These 'third-order problems' are likely to grow in urgency and include not only the compatability of 'new' life forms and existing ecosystems, but also the relationship between conserving species' diversity in laboratories, and gene banks, as well as in the wild.

STRUCTURE AND AGENCY IN THE MODERN FOOD SYSTEM: THE MANAGEMENT RESPONSE

The early part of this chapter addressed the question of agricultural policy instruments, and the way these had been used to support intensification, at the cost of environmental considerations. Given the high visibility of environmental issues and, in the developed countries, the acknowledgement that environmental considerations require changes in agricultural policy, it has become important to review the various

217

ways in which the rural environment can be more effectively 'managed'. Beginning with an analysis of environmental management in rural Britain, and extending this to the rest of the Community, this section goes on to assess the kinds of farm-level response which have been induced by recent policy making. The question posed by this discussion of sectoral and farm-level decision-making, is whether it carries implications for developing countries, where the penetration of the modern food system is also eliciting environmental management interventions. This means looking very carefully at the assumptions underlying environmental management itself.

As Potter (1989) observes, the increasing concern with long-term support for farming in Britain and the European Community generally has 'coincided', especially in the British case, with public concern about the environment and its conservation. As we have seen, policies which encourage agricultural intensification, specialization and the concentration of farm ownership in fewer hands have been called into question by the existence of farm surpluses and the heavy financial cost of supporting the agricultural sector. It needs to be emphasized, however, that it is public involvement in conservation, and the lobbying of environmental interest groups in Britain, which has called the model itself into question.

The agriculture and environment debate in the United Kingdom dates back to the early 1960s, although it did not reach the European stage until almost twenty years later (Potter 1989: 137). Beginning in the 1960s the public's attention was alerted to the cumulative effects of using pesticides in agriculture. The Nature Conservancy began to reveal that there had been dramatic losses in some wildlife species, such as peregrine falcons, due to the use of organo-chlorines (Hooper 1984). Within a decade this concern, fostered by the growth of the countryside lobby, had extended to the loss of whole habitats, such as moorland, heathland and ancient woodland, losses which were attributed to agricultural intensification and land reclamation.

Groups such as the Council for the Protection of Rural England (CPRE) which had considerable following, especially in the more scenically attractive parts of the British countryside, drew attention to these environmental losses. Other controversies were more long-standing but no less vituperative: the ploughing of moorland in the Exmoor National Park, for example, drew attention to the limited powers exercised over supposedly 'protected areas' (MacEwen and MacEwen 1982). Land use controversies began to engage public attention as much as environmental pollution or contamination from intensive farming practices. Furthermore, the effect of EC membership in accelerating these losses gave rise to additional concern at the environmental costs of 'productivist' agricultural policies.

The loss of the countryside's amenity value was a particularly British concern. As Potter (1989) shows, the agenda was partly set by relatively autonomous bodies known as 'quangos' within the structure of the British government, notably the Countryside Commission and the Nature Conservancy Council. In 1974 the Countryside Commission published its 'New Agricultural Landscapes' study, which documented the ways in which familiar features of the British countryside, such as hedges, copses and ponds, had been removed, as farmers deserted mixed husbandry for more specialized farming operations. By 1977 the Nature Conservancy Council (NCC), largely staffed by trained environmental scientists, with no obvious 'political axe' to grind, had reported that existing policies, which sought to protect sites of special environmental interest rather than more integrated land use practices, were seriously deficient (NCC 1977). By 1984 the NCC had revealed that only 3 per cent of lowland grassland had remained unaffected by agricultural intensification, since 1945, and that ancient lowland woodlands had been reduced by between 30 and 50 per cent in the same period (NCC 1984).

While the environmental lobby was beginning to raise its flag during the 1960s and 1970s, evidence was emerging that the agricultural lobby was being forced to adapt, slowly and painfully, to the realization that it could no longer count on the unqualified support of British governments. Two agricultural White Papers in the 1970s (MAFF 1975; 1979) departed, in significant ways, from the conventional wisdom that, as the major land-using industry, agriculture could expect to expand unchecked. Although no revised national policy existed which integrated the agricultural restructuring being urged on the European Community with more attention to environmental objectives, questions such as the appropriate size of the agricultural industry in terms of food and fibre output had become matters of policy debate. As a recent research paper makes clear, a subtle shift had occurred between 1975 and 1979, in which the Ministry of Agriculture conceded that the 'national interest' was distinguishable from that of farmers (ESRC 1989: 6).

By 1981 the tide of concern over the demise of the British countryside led a Conservative government to introduce a Wildlife and Countryside Act which provided a legal basis for management agreements between farmers and the NCC on farmed land of special scientific interest. This Act allowed for payments to be made to farmers, by way of compensation for foregoing modern agricultural practices, in return for agreed conservation practices. Owners and landowners were required to seek permission from the NCC if they intended to change their farming practices. The scheme attracted considerable criticism for being 'too little, too late', resting on the farmer's goodwill, and proving very costly to the Exchequer (Lowe *et al.* 1986; ESRC 1989). Although the shift away from unbridled productivism had occurred, the Government seemed anxious to proceed by agreement and co-operation with farmers, almost whatever the cost. Voluntary and informal structures of land use planning and regulation were very much the method favoured by the Government.

It is this concern with 'management agreements' that marks out recent policy interventions in the industrialized countries. According to a recent report from the OECD there are three desirable aspects of such agreements. First, the environmental benefits have to be realized effectively. For example, if the aim is to prevent cropping, then the cropping rights should be permanently acquired under the agreement. Second, the payment made in compensation for the reduction in expected net farm income, which the adoption of new management practices is likely to produce, should not exceed this 'opportunity cost'. Third, the agreements should not be in force long enough to enable farmers to bargain about the level of compensation, by refusing to operate under the new conditions until all their demands are met (OECD 1989). It will be clear from these 'desirable conditions' that it would be impracticable to introduce similar agreements on a large scale in most developing countries.

The mounting controversy about agricultural support, however, did begin to alter the policy debate by linking the redirection of agricultural support to rural conservation measures. Within the EC structures legislation (EC regulation 797/Article 19) and the enabling legislation contained in the 1986 Agriculture Act, financial aid began to be offered to appropriate farming practices as well as to farmers who refrained from using inappropriate practices. A number of areas considered 'environmentally sensitive' were outlined in the 1986 Act, and about 750,000 hectares of land were designated in the United Kingdom. Farmers were being paid, for the first time, to 'produce' countryside, something which, as we saw in Chapter 2, they did without significant agricultural support in the era before agricultural intensification. The principle that the interests of farmers were distinguishable from those of the British countryside was conceded in the 1986 Agriculture Act, which 'imposed a statutory duty on Agriculture Ministers to balance the conservation and promotion of the enjoyment of the countryside, the support of a stable and efficient agricultural industry and the economic and social interest of rural areas' (ESRC 1989: 6).

Although the agricultural lobby no longer commands undivided support, even from Conservative governments in Britain, and the environment lobby has increased in size and authority (Lowe *et al.* 1986), the mechanisms through which agricultural support can be shifted towards better conservation are still in their infancy. Writing about the United States, where the conflict is more often between agricultural intensification and sustainable agriculture, rather than a 'countryside' with amenity and recreational value, Buttel (1982) argues that farmers are most unlikely to change their practices to meet environmental guidelines:

> State policy-makers have been unable to call upon farmers to collectively and voluntarily reduce their production during phases of overproduction. Given the inability of farmers to control their production, state action to reduce production by rationing land resources has been undertaken because rationing other inputs (ie. petrochemicals) would be strongly resisted by farmers and by agribusiness firms.
>
> (Buttel 1982, 21)

Some commentators, such as Potter (1990) have argued that the preservation of existing farms is a necessary component of any wider conservation goals in European rural areas, and that the offering of income aids to farmers who agree to make their farming practices more environmentally sensitive is the most direct and powerful way of advancing future rural environmental policy. However, the same writer is equally aware of the limitations of such a small-farm approach. He writes that:

> realistically, though, there are limits to the extent to which a Farm Survival Policy can be orientated towards conservation objectives. With the notable exception of environmentally sensitive areas it is unlikely that the disadvantaged and intermediate farms that will be targeted in future will be located in areas of greatest environmental potential or vulnerability. The very transparency of expenditure on income support presents another difficulty: it may not be desirable to seek to achieve long-term environmental goals under expenditure headings that may be vulnerable to political pressure.
>
> (Potter 1990: 11)

It is not clear, then, that the situation in Europe is as strikingly different from that of North America as first appears. The powerful economic lobbies behind high-input farming have not been seriously affected by the new calls to conservation. Indeed one feature of so-called 'Green consumerism' today, as we shall see in Chapter 7, is that commercial interests with dubious environmental records are often loudest in their support for more 'sustainable' practices. The political climate has changed in recent years, and this is not an unimportant fact in itself, but the model which encouraged agricultural intensification as part of the development of the modern food system, has been dented rather than overturned. In the process, 'management' has acquired both greater credibility and a more uncertain future.

ENVIRONMENTAL MANAGEMENT AS A POLICY INSTRUMENT

André Gorz (1989) makes a distinction in his work between what he calls 'environmentalism' and 'ecologism'. Environmentalism refers to planned intervention to secure improvements in environmental quality, much of it undertaken by the state in developed countries. Ecologism represents much more deep-seated objections to the way the environment is handled in such societies: 'Ecologism is not content with correcting the environmental impact of the system. It examines the reasons *why* certain technologies and forms of production and consumption were promoted in the first place' (Gorz 1989: 4). In the developed countries, as we have seen, the state already intervenes to correct or modify the full effects of the development process on the environment. In such countries, the 'core countries' for the modern food system, the popular mass diet is already well established, and agriculture employs only a small proportion of the population.

Under these circumstances, managing the wider environmental implications of the food system assumes importance largely because basic needs have been met. Environmental policy interventions usually take two forms as far as food system activities are concerned: the protection of 'valued' environments from the pressures of human economic activity and the reduction of market pressures on specified groups of agricultural producers, to enable them to act more 'sustainably'. Both kinds of intervention are, arguably, holding devices, since they can be continuously negotiated and renegotiated. In fact the more interventionist the state becomes the larger and more vocal the environmental lobby usually becomes, exposing new environmental risks and dangers, to which it is claimed government should be responsive. The principal objectives of environmental policy, in relation to agriculture, speak to the priorities and concerns of urban middle-class groups: amenity recreation, species preservation and aesthetics (see Table 6.2). The membership of environmentalist organizations has increased rapidly in recent years (Table 6.3). In developed countries the existence of protected areas does not diminish opposition to environmental management decisions, but it does draw attention to some of the problems and limitations of what we have described as the 'modern food system' model.

In developing countries the situation is usually markedly different. Environmental conflicts in developing countries are about poor peoples' access to the resource base in the struggle to maintain livelihoods, and the conflicts which are associated

Table 6.2 UK countryside recreation 1985

Activity	Percentage of people (aged 12 or over) taking part in each activity during 1984
Drives, outings, picnics	55
Visits to historic buildings	39
Long walks	33
Visits to country parks	27
'Pick your own' (PYO) farms	23
Nature reserves	13

Source: Countryside Commission, National Countryside Recreation Survey, *Social Trends* 17 (1987)

Note: In any four week period during 1984, 40 per cent of the population made no visits to the countryside, while 17 per cent made nine visits or more.

Table 6.3 Membership of environmental organizations in the United States (thousands)

Organization	1970	1975	1980	1985	1990
National Wildlife Federation	2,600	n.a.	4,600	4,500	5,800
Sierra Club	114	153	182	363	566
National Audubon Society	105	255	310	425	515
Wilderness Society	66	n.a.	63	97	363
Environmental Defence Fund	10	40	45	50	150
Greenpeace*	–	6	80	450	2,000
National Resources Defence Council**	–	15	35	65	140

Source: *The Economist* 21 April 1990

Notes: *founded 1971
            **founded 1970
            n.a. not available

with changes in resource use in the course of economic development. 'Sustainability' in this context is not primarily a consumption goal of the relatively leisured middle class, but a concept which, if it is espoused by anyone, is espoused by the rural poor themselves and usually within the context of human rights. Of course, this is not always the case, and the fact that the rural poor often stand in an ambivalent position to more sustainable systems of agricultural production is evidence of the difficulty in securing livelihood advances.

It is worth emphasizing, too, that the interest of people in developing countries in their own environments does not put a seal on environmental conflict in these societies. There are also issues of the 'international commons': common property resources, as well as open access resources to which no property rights apply. Many of the world's life-support functions are dependent on the protection of resources in large tracts of the South; genetic materials provide the best example, as we have seen, but increasingly even issues like rain forests, climatic change and the disposal of toxic wastes involve relations between the developed and developing countries. The globalization of the modern food system ensures that changes in resource use, and the environmental effects of new technologies, are transmitted from one geographical location to another, just as patterns of food consumption have become increasingly international. Policy and management interventions to secure more 'sustainable systems' cannot be judged simply in terms of their own logic and internal dynamic; they need to be considered within the broader context of production, consumption and the transformation of resources which defines the modern international food system.

As the British response to the World Conservation Strategy (1983) made clear, 'any useful analysis of the reasons for natural reserve destruction would have to include questions of wealth distribution *within* and *between* nations' (World Conservation Strategy 1983: 324). The point about environmental management, as it is practised in the developed and developing countries alike, is that it takes as a 'given' the distributive consequences which the market produces in the course of development. In developed countries, as we have seen, increasingly vocal and well-organized groups have mobilized around environmental objectives. In developing countries only the poorest and most marginal are likely to do so: rubber tappers in the Brazilian Amazon, Chipko activists in India, women farmers in the Kenyan 'Green Belt' movement. The politics of environmental movements, as well as management interventions, have a different basis in the South:

The more aid flows towards those sectors [in developing countries] favoured by political, industrial and commercial criteria, the less likely they are to arrest such environmental damage as is occuring. This is because the greatest environmental damage, and needs, tend to occur in the rural areas and amongst the urban poor of the least developed countries (with the obvious exception of industrial pollution which is not insignificant in some developing countries).

(World Conservation Strategy 1983: 328)

These distinctions between the role of environmental management in the agriculture of developed and developing countries are illustrated in Table 6.4. In the developed countries the management of the rural environment is contested by two opposing interests: those of farmers and conservation bodies. The environmental lobby has assumed increasing importance in the United Kingdom because environmental values are located in both resource-rich and resource-poor areas. Although some relatively remote areas are underpopulated and offer few attractions to agriculture, most rural areas are within reach of heavily populated urban settlements and land uses are strongly contested.

In North America, Australia and parts of Europe the environment presents itself in another form, that of wilderness, which has served as a focus for the environmental lobby, both as intellectual inspiration ('deep ecology') and as a barometer of cultural, and hence political, concern. Nash (1967), in *Wilderness and the American Mind*, called these areas of wilderness 'institutional containers', vital because they have 'resisted' civilization.

The juxtaposition of dense urban populations and resources worthy of conservation, in the developed countries, enhanced the importance of 'protected area' status, such as that of National Parks. Most of these areas had little agricultural potential, but enormous cultural and political weight was attached to their conservation. The difficulties afflicting areas of some agricultural importance, which contained or bordered zones of environmental value, were rather different. Strategies have had to be evolved for reducing productivist pressures in these areas, either through 'setting aside' land deemed surplus to agricultural requirements or by shifting towards 'lower-input' farming. Beginning with physical planning controls, and protection of areas designated as being of special interest, the move within the countries of the European Community has been towards more economic incentives to practice 'sustainable agriculture', including measures to support farm families as

Table 6.4 Agriculture and the environment: an international perspective

	Conflicting interests	Location	Policy measures	Outcome
Developed countries (EC)	Farmers versus national conservation bodies	(a) Resource-rich areas (e.g. wetlands)	1 Protected area status 2 Set-aside	(a) Reduce level of agricultural support
		(b) Resource-poor areas (e.g. uplands)	3 Reduced input subsidies 4 Income support to farmers to practice sustainability	(b) Planning controls State mediation: • beneficiaries chosen on resource quality and location • environmental budget linked to lobby and agricultural support costs
Developing countries	Rural poor versus rural rich	(a) Protected areas (e.g. common property resources)	1 Protected area status 2 Environment-friendly technology encouraged (FSR)	Conflicts intensify between national governments and international agencies
		(b) Other areas (by default)	3 Debt/Environment swaps	(a) priority of environmental and development objectives (b) 'sovereignty' of national resources at issue (c) funding for environmental policy objectives

well as farmed landscapes. The motivation to undertake more management interventions has been provided not only by heightened environmental concern, but the onerous budgetary costs of supporting agriculture and production of food surpluses. The column on the far right of the table 6.4 indicates the degree of discretion available to governments in the developed countries over the financing and 'targeting' of beneficiaries in environmental policy interventions.

The comparison with the developing countries makes it clear that environmental management principles cannot simply be transferred from North to South. In developing countries the contest over natural resources, their use and ownership, is between the rich and the poor in rural areas. In most cases environmental agencies are influenced by external pressures – such as those associated with opposition to tropical deforestation – and, in any case, such agencies have no 'teeth'. Protected areas such as Biosphere Reserves are demarcated, but the means with which to ensure their protection are rarely available; budgets are minimal, 'policing' activities random or unenforceable. In areas other than protected areas there is little, if any, outside intervention to protect the environment. In some cases, agricultural technology is disseminated which is more 'environment-friendly', such as that of Farming Systems Research (FSR), but this technology is encouraged largely because high-input agriculture is impossible on marginal or resource-poor land. Only relatively recently have 'sustainable' objectives begun to be incorporated in production programmes for small farmers, but the vast majority of the rural population, particularly rural wage labourers, lie outside the reach of existing 'sustainable' policy packages. It is extremely unlikely that policies like the 'polluter pays' principle (Pearce *et al.* 1989) could be incorporated within the gamut of policies influencing rural resource uses in developing countries.

The pressures mounted within the international community to call a halt to environmental degradation in the South have had little impact on the economic policies and practices that largely govern North–South relations. One much-debated exception is that of 'debt-for-nature' swaps, in which some areas of tropical forests are 'protected' from development, and their value written-off against part of the country's external debt. In practice, however, such swaps feature in only tiny areas of endangered land, it is unlikely that they can be adequately protected, and in any case the decision to do so was not arrived at democratically but by administrative fiat. We have yet to see even the beginning of effective local participation, by indigeous communites and others, in the long-term sustainable management of tropical forest areas. What has emerged from the 'debt-for-nature' debate is a measure of hostility between some governments in the South, anxious to protect their sovereignty and forced to increase foreign exchange earnings by selling products that carry environmental costs, and what are sometimes perceived as 'interfering' First World environmentalists. Developing country governments argue, not unreasonably, that they cannot be expected to practise sustainable development when their paymasters, the developed countries, fail to do so.

ENVIRONMENTAL MOVEMENTS AND SUSTAINABLE DEVELOPMENT

Most of the recent discussion of 'environmental management' in developing countries has paid scant attention to the role of local people (Schramm and Warford 1989; Barbier 1989). Environmental management is defined almost exclusively terms of a rigidly 'top-down' approach. It is assumed that by incorporating better systems of environmental accounting in the national accounts, and by ensuring that the negative costs of development are allowed for within project budgets, progress will be made towards better management of the environment. Desirable as it is that economics takes environmental problems seriously, such measures are unlikely to affect the balance of forces which cause poor rural people to place excessive strain on their environments. The 'official' discussion of environmental management in the South makes barely a reference to the increasingly important experiences, and literature, on environmental movements there (Hecht and Cockburn 1989; Guha 1989; Shiva 1989). The remaining section of this chapter focuses attention upon the need to re-examine our knowledge and understanding of the environment, in the light of the experiences of other cultures far removed, socially and geographically, from the 'core' of the modern food system.

SUSTAINABILITY AND DEVELOPMENT

It is clear that rural development projects have often neglected local, ecological and cultural factors in the South. It is also clear that structural factors – which influence people in not acting 'sustainably' – have frequently been ignored (Blaikie 1985; Redclift 1984). We need, in fact, to take account of both the cultural and structural dimensions in any definition of sustainability. Then we can 'begin to ask whether there are patterns of myth and social organisation, which are more likely to ensure and maintain man's [sic] survival base' (Burch 1983: 307). It is clear that environmental problems are not new. Historical, and archaeological, evidence suggests that early societies often behaved in ways which were damaging to their environment (Hughes 1975). These societies also had to 'solve the basic problem of producing food surpluses and collecting raw materials from rural areas to sustain large urban populations' (Weiskel 1989), a problem that faces countries in the South today. However, such societies, unlike the South today, were not faced by twin assaults from rapid economic and demographic growth, on such a scale. They did not exist within a global context which placed their own food systems and values in jeopardy.

If we are to understand fully the relationship between sustainability and development, we must begin by recognizing that they are paradigms with historical as well as geographical antecedents which need to be located within specific social and economic contexts. The point at issue is that societies differ in the way they prescribe courses of action towards the environment. Too often we assume systems possess universalistic characteristics, including the food system. Such universalistic approaches to social and political systems often leave no room for human agency within the framework of explanation. The question we might address in exploring environmental movements was posed by Giddens in the following way: how does the 'practical consciousness' that people acquire from their society, and from their environment, relate to the 'structural properties of the system' (Giddens 1976)? Environmental movements include movements of resistance to the encompassing food system. Under what circumstances are the structural properties of such a system called into question?

The epistemological dimension

To answer this question we might begin, not with an analysis of political conflict over the environment, but with the way in which the 'environment' is understood. Instead of working 'backwards' from social movements to consciousness, we can work 'forwards', from the formation of environmental consciousness to the organization of alternative forms of social action to redress environmental problems.

It is clear that the way in which knowledge is gained about the environment and the way in which this knowledge is used have considerable bearing on the question of human agency, and with it environmental management practices. Howes has argued that in agricultural systems engaged in monocrop cultivation the world view is one in which 'society is conceived as an entity apart from Nature' (Howes 1979: 23). It is clear, too, that many non-Western cultures possessed 'scientific theory' before Western science, as we know it, had fully developed. Looking at examples from South Asia, Goonatilake claims that, 'their large-order cognitive maps anticipated the non-manipulable, and distant reality of modern science' (Goonatilake 1984: 3). Furthermore, these kinds of knowledge system often codify knowledge in ways that ensure its cultural survival, long after the reasons for certain observances have passed. In some cases 'after an initial map has been made and legitimized . . . it tends to continue to be believed for considerable periods of time' (Goonatilake 1984: 3). A useful distinction, although it should be emphasized that it exists *within* societies as well as between them, is between what Feyerabend (1987) calls *historical traditions* and *abstract traditions*. The two traditions can coexist in a society at any point in time, but some cultures give more emphasis to one tradition than to the other.

Historical traditions involve a different discourse from abstract traditions. It is assumed that 'the objects already have a language of their own' and it is the task of people within a society to learn this language. Their task is to 'learn the language of the objects as they are, and not as they appear, after they have been subjected to standardizing procedures' and experimentation (Feyerabend 1987: 115). According to this tradition there is no such thing as pursuing 'objective truth',

231

since truth involves both an object and a subject. Feyerabend compares this with what he terms an abstract tradition.

Abstract traditions are different in that they involve framing statements as a means of building up a model of what happens in nature. According to Feyerabend, 'the statements are subjected to certain rules (of logic, testing and argument) and events affect the statements only in accordance with the rules' (Feyerabend 1987: 116). It is possible, then, to make scientific statements without personal contact with the objects being described. Both these epistemological traditions exist, at different levels, in most societies, but within the Western scientific tradition much more emphasis is given to abstract traditions. In this respect the development of Western science is best understood in the context of Judaeo-Christian thinking, which has often appeared inimical to 'scientific' method. For example, Goonatilake (1984) argues that religious and philosophical objections to scientific method in Europe would not have occurred in South Asia, where most religious beliefs emphasize the role of both abstract and historical traditions in arriving at valid statements about the world around us. The first point that needs to be emphasized in establishing a framework for understanding environmental conflict in the South is that a society's epistemological framework will influence its understanding of the environment. What we perceive as 'environmental' problems may be interpreted in other cultures as essentially due to the breakdown of social and cultural practices and beliefs.

In addressing these questions we need to be aware that our understanding of societies very unlike those of the industrialized countries frequently distorts our understanding of their environments. By the same token, the rediscovery that knowledge-systems other than our own have importance for understanding our relationship to the environment has become a central, but contentious, tenet of recent Green thinking (Capra 1976; Illich 1975). As Luke (1988) observes, some Green thinkers have misunderstood the importance of non-Western science. 'Deep ecologists' (Devall and Sessions 1985, for example) have sometimes equated small-scale, non-Western societies with 'primal societies', implying that 'primitives' provided an echo of our own pre-Enlightenment past. This rather naïve view of the superiority of non-Western epistemologies fails to recognize that myths, magic and ritual are 'the functional equivalents (and perhaps the conceptual antecedents) of Enlightenment science and technology' (Luke 1988: 76). This kind of simplification and misunderstanding also errs in lifting ecophilosophies out of their cultural context – Taoism or Buddhism – 'with little consideration of any cultural grounding' (Luke 1988: 78), when it is the specific cultural context of environmental epistemologies that gives them their force and vitality. As an approach, the close association of indigenous knowledge with 'answers' to current environmental problems is frequently ethnocentric, ahistorical and reductionist, substituting descriptions of the world for more rigorous and less ethnocentric analysis.

Other examples exist of societies closer to our own in which environmental problems are constructed from a very different political and social milieux. In Eastern Europe today, for example, the commitment to radical political ecology, and the growth of Green movements, is also linked to processes that are not strictly 'environmental' in the narrow sense employed in the West (Redclift 1989). Conflicts over ethnic identity, and the struggles for civil and human rights in contemporary Eastern Europe, have incorporated – and adapted to – environmental demands, and often express these demands in language drawn from the discourse.

The economic dimension

The second, important dimension of relations between the environment and society is the economic dimension. The environment is subjected to structural changes over time, linked to economic growth and the development process. If we confine ourselves to the relationship between agriculture and the environment we can identify several features of the development process which have major implications for environmental systems, among them: the production of food as a commodity rather than for household consumption; the adoption of 'high-technology' energy systems on the farm, and in the food processing industries, rather than 'low-technology' energy inputs; and the role of technology in reducing the pressure exercised by population increase on limited natural resources. These factors will be looked at in turn.

Food as an exchange value

In some societies agricultural practices underwrite sustain-ability, in the sense that consideration of the long-term impli-cations of current agricultural practices is enshrined in farm management practices. For example, Wilken (1987) shows how some small farmers in Central America, without recourse to fossil fuels, effectively 'bank' their own labour, and that of their household, investing labour in resource management practices that make sense within a longer time-frame. They are seeking to reduce future operational costs on the farm by making major current investments in labour-intensive management now.

As we have argued above, in some societies such practices become ritually encoded. Rappaport (1967) shows how, under stable conditions, the ritual enactment of myths may not only help to maintain the social cohesion of the group (a point often made by anthropologists) but also help to sustain the natural environment. Of course, the value placed on a resource is linked to other social referents, such as the concept of owner-ship, as well as economic factors like the cost of extraction. The development of wider markets for food products, and the adoption of production techniques that depend on sophisticated off-farm processing and industrial inputs, does not necessarily mark the point at which food becomes a com-modity – in most societies food has been a commodity for a long time. However, it does mark the point at which the con-version of food from a use value to an exchange value has the greatest impact on the way the environment is managed, and ultimately socially constructed. Food is taken from the province of reproduction in small-scale societies and transferred to that of production and consumption in more complex societies. Eventually, in what we can term 'post-industrial societies', the locus of dissent shifts again, this time from production rela-tions to those of consumption. Some of the most important implications for the evolution of the food system are represented by resistance to the introduction of new, processed 'convenience' foods. This is discussed in Chapter 7.

Food and energy transfers

The food system also affords an excellent example of the way that energy use is linked to the capacity of ecosystems to retain, as well as transfer, value. The process of development, as we have seen, is usually considered in relation to the movement of capital and labour, rather than the exploitation of natural resources. The structural and spatial aspects of transfers of value from one part of the (developing) world to another part of the developed world have been emphasized by O'Connor (1988) and Smith (1984). Without dissenting from the significance of these forms of surplus extraction, it is important to emphasize that transfers of energy from within one system to another can also be seen as transfers of value. Bunker (1985) uses both ecological principles, and the laws of thermodynamics, to assert that 'our calculus of value must include not just the labour and capital incorporated into commodities, but all forms of energy and matter'. In his view, 'the embodiment of energy in economic and social organization encompasses far more of the essential differences, and relations, between core and periphery, than measures linked to commodity production and exchange can' (Bunker 1985: 239).

Population increase and carrying capacity

Economics has also experienced considerable difficulty recently in incorporating the full implications of changes in environmental systems within its analytical framework (Pearce *et al.* 1989; Barbier 1989). The Malthusian tradition emphasized that population pressure placed limits on the capacity of natural resources to feed the population. In fact, like many economists, Malthus did not have a great deal to say about the environment. Barbier (1989) reminds us that the pessimism of the Malthusian tradition was only matched by the optimism of the Ricardian view of scarcity. This placed emphasis on 'relative scarcity' rather than 'absolute scarcity'; as resources became used up substitutes were found for them, extending the time-frame in which the environment could facilitate economic growth. The problem of carrying capacity was not solved but it was 'put back', and technology was the major factor in effecting this change.

Later neo-classical thinking, represented by Alfred Marshall, took an even more optimistic position than Ricardo; no tendency was predicted for profits, wages and rent to decline with increased scarcity of natural resources. The implication was that we were entering an 'age of substitution', the environment could be exploited without degradation being inevitable, indeed the scarcity of non-renewable resources might attract a higher price, and therefore lead to their better conservation (Barbier 1989). This view of 'substitution' still plays a large part in the way we conceptualize the environment today, and has been given added impetus by developments in biotechnology, as we saw in Chapter 5.

The importance of the economic dimension is that it necessarily involves considering the conversion of natural capital into social capital, economic processes of growth and development do not simply modify, they actually transform the environment. In the course of these transformations the development process 'hands back' to nature some serious problems, including that of sustaining economic levels of production and consumption.

The political dimension

The development of the modern food system has brought new urgency to another important dimension: the political dimension. The wider parameters of the food system have introduced this dimension into social relations on the farm, relations which were previously mediated by more endogenous factors. Linkages between agricultural producers, suppliers and markets have necessitated more joint decision making, and have led farmers and the food industry into political alliances. As producers and consumers have become incorporated into wider market systems, their response to their tacit incorporation has often taken the form of resistance, in which class, gender or ethnic subordination is contested. The 'fields of resistance' that result have often served as rallying points for disaffected groups with different political agendas. The concept of fields of resistance we take from Foucault (Smart 1985). It is used to distinguish between different responses to the 'individualising techniques of power', through which systems are able to exert social control and often develop a political consensus. The individual's response to the management of this consensus lies at the heart of the resistance to industrially produced food in the developed countries. Attempts to create a consensus around modern agriculture and food policy have recently broken down, as we shall see in Chapter 7.

236

Among the forms of resistance which environmental movements express we can include:

(a) *exploitation*, through which the individual is controlled via market forces and the production process;
(b) *domination*, through which the individual's identity is defined in terms of gender or ethnic roles; and
(c) *subjection*, where the controlling effect of dominant belief systems in the society tie the individual to an ideological position.

An analytical framework based on fields of resistance is of use in exploring the political changes that accompany the development of the modern food system, and the transformation of natural resources. For example, in the case of farmers in the United Kingdom the growth of market forces and production relations on the farm proved to be the decisive factor, at the turn of this century, in bringing about national organizations of both farmers and farm workers from different class positions. However today, farmers and landowners are united in their resistance to restructuring brought about by the market, as well as to government policies on conservation or 'countryside management'. Farmers have become dependent on the state, and at the same time resentful of their dependence – an ambiguity behind the attraction of agrarian ideology, especially in countries like France and Germany, where farms were smaller and the dislike of state corporatism more pervasive. By contrast, the importance of *subjection*, as a field of resistance, has grown with the increasing effect of modern agricultural practices on the environment. The demand for more 'organic' food, the interest in preserving the physical condition of the countryside, and political mobilization around issues like animal rights and food quality have brought a new urgency to the interface between agriculture and the environment.

The political dimension exists in any set of environment–society relations. However, we need to ask why resistance sometimes finds expression through collective activity (land-rights, urban struggles, etc.) and why it sometimes implies withdrawing from the wider public policy sphere into an 'alternative' community or individual 'survivalist' position, at odds with the mainstream political system. These questions inevitably raise wider issues concerning the conflicting uses to which the environment can be put. The conflicts that surround resistance to 'development' bring into the open alternative strategies for utilizing local knowledge as well as conserving nature. The analysis, at this point, cannot be confined to the food system, and embraces the trajectory under which autochthonous alternatives to development diverge from mainstream economic forces under the direction of capital located in the North. These wider issues are discussed at length in Redclift (1991).

Having considered the principal features of the modern food system from a historical and cross-disciplinary perspective, it is necessary to take the analysis forward into the realm of contemporary culture and politics. This involves analysing the growth of opposition to changes in the food we consume. This chapter has been principally concerned with exploring the effect of the developing food system on rural land use. Increasingly, the focus of concern, for interest groups, politicians and the consumer, has shifted to what we eat, and changes in our diet in turn reflect wider changes in consumption and the use of leisure, as pointed out in Chapter 1. The food system assumes its most radical and yet most familiar form in the supermarkets where we in the developed countries buy most of our food. Resistance to changes in food marketing, processing and retailing have assumed a greater importance in recent years. Today they represent a veritable counter revolution.

7

CONCLUSION: COUNTER REVOLUTION

At the beginning of his novel, *The Mezzanine*, Nicholson Baker's central character visits a small food chain outlet in New York, 'Papa Gino's', where he buys a carton of milk, and is offered a straw by the counter assistant, called 'Donna', according to her name tag. He pauses, considers the question with the kind of seriousness that people usually reserve for questions of food and drink, and decides that the answer (psycho-parenthetically!) deserves a footnote. There follows a long, rambling and utterly absorbing footnote, dedicated to the question of why and when 'all the major straw vendors switched from paper to plastic straws, and we entered that uncomfortable era of the floating straw' (Baker 1988: 4). In the course of this footnoted journey Baker touches on several facets of the consumption end of the modern food system that reach into our private lives, become ritually embodied in daily life, create fixations and aversions, and ultimately correspond with major shifts in food manufacturing, human labour and recreation. It is a tribute to the elegance of the author's prose that this journey is undertaken without ceasing to take the subject seriously: we may seek to trivialize our dependence on fast foods, but fast foods are not in themselves trivial.

This chapter examines some of questions surrounding the expression of consumer preferences for food, and the way they have been engineered, and orchestrated, by the food industry, shifting attention from the production of food and the management of the natural environment to the consumption of food, in the home and behind the retail counter. This involves attention to a number of issues which have already emerged at various stages in this book: patterns of resistance to the industrial food system; definitions of nature and what constitutes 'natural'; the phenomenology of food consumption; and the contribution of political economy to the analysis of the food system. Finally, we draw together the different threads of our analysis, from different sections of the book, and examine the implications of the development of the modern food system for countries in the South, as well as those in the industrialized North.

FOOD CONSUMPTION: PATTERNS OF RESISTANCE

Recent studies in the United States, notably that of Belasco (1989), have drawn attention to the way in which changes in the food industry have provoked resistance from some sections of consumers, and the demand that government intervene more to protect the interests of the food consumer. The food industry, in turn, has counter-attacked in a number of ways to the barrage of criticism and public consciousness surrounding food issues. Belasco details these 'counter-revolutionary' (or perhaps 'counter-counter-revolutionary' would be more accurate!) strategies: food manufacturers have sought to dilute recommendations ('reduce' rather than 'avoid' foods that are bad for you); they have employed what marketing people call 'repositioning', altering the way products are described rather than the products themselves; and they have increasingly employed strategies both to differentiate products and to differentiate markets.

In the United Kingdom today we could scarcely fail to be aware of the importance of product differentiation: the food industry has employed advertising techniques to differentiate between products for many years. However, what is new, and very much a response to consumer resistance, is product differentiation in 'healthy' or 'ecologically-friendly' foods, an important aspect of what has been dubbed 'Green consumerism'. Similarly, the extent to which the market can be differentiated according to the tastes of different groups of consumers, has received widespread attention, notably in the food trade journals (Mayo and Denison-Ender 1988, Dade 1988). We would argue, in fact, that some segments of the market in developed countries have entered a 'post-fordist' era, as far as the consumption of food is concerned. That is, an era not so much of the mass consumption of homogeneous, industrially produced food as of the development of healthy foods for market 'niches', often reflecting ethnic variety and traditions, but utilizing the full armoury of the food processing industry, and targeted to consumers willing to pay for high value-added products. With the growing sophistication in manipulating the chemical constituents of food, highly processed food can be presented as 'healthy' food, manufactured to have low fat, low cholesterol and high fibre content, in line with current dietary recommendations.

The challenge facing the food industry, and this is linked to shifts in the food system more generally, is how to make healthy foods more marketable, and marketable foods appear more healthy. Both questions are capable of eliciting a response from food manufacturers, but to understand the background to this response we need to examine the patterns of resistance which have grown up to recent changes in the food industry, patterns which echo some of the forms of resistance to environmental changes discussed at the end of the previous chapter.

The beginnings of what Belasco calls a 'countercuisine' are linked to a number of quite important social changes which, as we have already noted, shift the individual's attention from the sphere of production relations to those of consumption. Three aspects require our attention: the way in which we arrive at our impressions of what foods to avoid and what foods to consume (the quality of food as a commodity in the market); the way in which food 'signifies' and expresses other aspects of our lives (food as self-enhancement); and finally, the way in which our consumption of food is linked to broader questions of ownership and organization in the economy as a whole (food as part of a broader political economy). Almost unnoticed, issues surrounding each of these aspects of the consumption of food have assumed enormous importance in recent years, but close examination reveals that what at first appear to be almost random evidence of social resistance to the modern food system have their roots in fundamental social changes. These social changes need to be outlined, and their implications examined. It is our contention that too much attention has been devoted to the 'signifying' aspects of food and diet, the semiology of food consumption, by sociologists and anthropologists, to the detriment of a wider understanding of the social transformations implied by the economic and technological changes in the food system.

As we have seen, especially in Chapters 1 and 2, women's roles have been transformed in a number of ways, some of which are clearly irreversible. Domestic work, including housework, has been opened up to important market forces, and the manufacture and purchase of 'white goods' for the home is now an important activity. At the same time the idea that housework is no longer onerous has served the interests of manufacturers and others. It is far from clear that technology in the home is labour-saving; this assumption, as we saw in Chapter 1, needs to be examined critically. Evidence exists that a human price is exacted for the benefits of speed and convenience which 'white goods' bring. Housework standards change, and women expect (and are expected) to raise the standards of housework in turn. At the same time, it is likely that women's release from the worst forms of labour drudgery is associated with more time being spent by them on other activities in the home, particularly attention to children. This is not to say that many household tasks are not easier; they are. It is merely to point out that the definition of housework, and with it women's responsibilities in the home, is constantly shifting. It is not immutable.

It is also clear that since the 1970s important changes have occurred in popular understanding of the relationship between food and health. Preventative medicine may only be in its infancy, and receive little official encouragement, but for some groups of people in the industrialized world, healthy eating is now considered essential. Healthy eating in the past depended critically on local custom and diet and, most importantly, income. Today this is still true, but local variations have less importance, and additional factors have made their appearance, which have engaged the attention of a battery of professionals, such as health educators, doctors, consumer groups and alternative therapists. The perceived need to diet to reduce obesity has given rise to a huge consumer market in low-calorie foods, including the controversial 'low-cal' liquid diets and innumerable diet 'systems'.

The publication of evidence about heart disease and cancer

has led to some marked shifts in the public consumption of certain foods, especially in the United States, where the changes date back to the publication of the McGovern Report in 1977. Food has become one of the obsessions of our time, associated as it is with good health, longevity and the occurence of stress. Physical fitness has also become an obsession in some quarters, perhaps also a 'fad'. Evidence from the United States suggests that, while in 1960 only a quarter of adults exercised regularly, by 1980 over half did so. The point is not that everybody is exercising more, but that more people are, and that health is increasingly linked to diet in many people's minds. This is not a trend likely to be discouraged by the food industry, since it also represents a change with commercial possibilities. Indeed, the giant food firms have jumped on this bandwagon, adapting to each new nutritional recommenda-tion – low sodium, low saturated fats, high fibre – to the point of confusing consumers grappling with labels and elusive media-speak, such as 'lite' and 'natural'.

Changes have also occurred in public perceptions of the two processes we referred to in Chapter 3: appropriation and substitution. Following the scare over DDT between 1969 and 1972 in the United States, and subsequent publicity about the use of other pesticides, attention has been given to the effects of pesticide residues in food, especially by the London Food Commission in the United Kingdom. Few studies have been undertaken, and even fewer given publicity, about the relative nutritional quality of organically produced and chemically produced food. Pesticide use in the United States rose 500 per cent between 1950 and 1986, but in the latter year one-fifth of United States's crops were lost to pests, the *same percentage as in 1950*! The tactic employed by the food industry in meeting the criticism of environmentalists and wholefood enthusiasts, has been that the industry can 'compensate' through industrial processes for the shortcomings of nature.

> elaborating on [the] ancient case for better living through chemistry, the modern-day Prometheans of the food industry spelled out the ways in which Mother Nature grossly neglected her human offspring, especially in the field, at the cash register and in the kitchen.
>
> (Belasco 1989: 117)

The industry argued that organic foods were more expensive (ignoring the hidden subsidies provided to chemically produced food), that organic foods could not meet market demand, that food additives increased the palatability of food, and that convenience removed the drudgery from food preparation. By the late 1960s in North America, and perhaps a decade later in the United Kingdom, food manufacturers began to 'reposition' themselves, as well as their food products. Now they were on the side of emancipated women and environmentalists; they offerred themselves as accomplices in women's drive for more independence and in the vanguard of responsible, sustainable resource management. Market researchers have credited much of the improvement in the food industry to their own responsiveness to consumer pressure. In the words of an industry spokesperson:

> For example, ten years ago the only place people could find additive-free *natural* products was in the comparatively cramped and small premises of health food stores. Today such products, not only much improved and – dare we say it? – engineered to satisfy consumer preferences, are mainstream products in prime placings in major supermarket chains the length and breadth of Britain.
>
> (Dade 1988: 108)

This represents one side of the story, but in practice healthier eating has also presented a challenge to the food industry which it has found difficult to grasp. The problem, in a nutshell, is how to leave out food additives and processing, and still add value to the final product.

The post-war period had seen additives 'substitute' for natural ingredients, while much that was natural had been removed from food. Food was given more value added through packaging, processing and ensuring it stayed 'fresh' longer. The challenge represented by making 'healthy' foods from 'health' foods was how to make profits by appearing to do less to the product. The industry responded to the challenge, as we have noted, by both market and product differentiation – identifying, and exploiting market opportunities for 'niche' products, such as low-calorie foods, ethnic health foods and fitness-related foods. Finally, the food industry has been in the forefront of the promotion of food supplements; people can now supplement their inadequate diet from a range of commercially promoted food 'accessories'. In 1985 over US$3 billion was spent in the United States on vitamin and mineral supplements and over $4.5 billion in supplements to fortified breakfast cereals. This diet supplementation was the inevitable outcome of the way eating had changed in the United States. In 1909, 40 per cent of calories had been provided from fruit, vegetables and grains. By 1976 only 20 per cent of calories came from these sources, the rest from fats and refined sugars (Belasco 1989: 149). As people ate more calories they ate fewer vitamins and minerals – supplements have helped to fill the vacuum left by changes in the 'affluent diet' of North America.

Another important social change that has accompanied the shifts in diet referred to above is that governments in the industrialized countries have been forced to make some accommodation to the pressures put upon them by food consumers, and interest groups representing consumers. In the United Kingdom this has come about largely as a result of successive food 'scares', such as those over salmonella, listeria and bovine spongiform encephalopathy ('mad cow disease') in the last few years. The extent to which governments will respond to consumer anxieties is still unclear, but the current debate about public nutrition policy and food legislation represents an excellent illustration of the kind of contradiction to emerge from the development of the modern food system.

The demand for more intervention in the food system presents difficulties for governments committed to the idea of choice in the marketplace. On the one hand it is difficult for them to resist, as a matter of principle, the demand for more information about food products. Issuing 'guidelines' about food and health is not inconsistent with conservative ideology, which places individual choice before government controls. The publicity given to the McGovern Report in the United States after 1977 served to divide the food lobby. Part of the food industry could take comfort from the encouragement being given to people to eat more cereals, vegetables and fruit. Another section of the industry tried to develop marketing strategies to overcome opposition to increased consumption of other items (particularly meat, dairy products and eggs). The development of a 'fordist' diet in the 1950s and 1960s had meant that red meat consumption rose, in the United States from 121 lbs per capita in 1930 to 160 lbs per capita in 1970. By 1968 20 million tons of vegetable protein was being fed to livestock to produce 2 million tons of animal protein. Since 1980 there has been a steady reversal of this trend: beef consumption has fallen in the United States, and voluntary reductions in food consumption, like those recommended by the McGovern Commission for eggs, sugar and animal fats, have also begun to be registered. However, this, too, is only part of the total picture.

Changes in diet, encouraged by public debate, have not yet pushed or persuaded governments, in the United States or Western Europe, to take a more interventionist stand on other aspects of food consumption, notably the use of food additives, although recent legislation like that proposed in the United Kingdom in March 1990 would introduce more rigorous standards of hygiene in the food industry. In the United States the Federal Food and Drug Administration (FDA) co-operates very closely with the food industry. In practice many of its recommendations are welcomed by the industry because they reinforce the idea that most food already on the supermarket shelves is fit for public consumption. The FDA-initiated testing of food led to the conclusion that over a thousand substances used as food additives were 'generally recognized as safe' (GRAS) for use in food, whereas only two substances were banned as a result of FDA testing: a carcinogenic weedkiller and, in 1969, cyclamates. The United States Government has proved reluctant to act as a watchdog on food additives. Meanwhile over 1 billion lbs of additives are added to food products in the United States, equivalent to 5 lbs of additives per capita (Belasco 1989).

THE PHENOMENOLOGY OF FOOD AND DEFINITIONS OF 'NATURE'

It is important to consider another aspect of food consumption: the meaning people attach to it, and the way in which food consumption practices help structure social relations more generally. Gofton (1989) has made a useful first attempt to come to grips with some aspects of the sociology of food consumption in the United Kingdom today. He notes several of the trends referred to above and in other chapters of this book: the growing concern for health; home-centred living; redefinitions of social roles within households; the trade-off between the time taken in food preparation and leisure activities; and the emphasis given in our society to individualism and hedonism (Gofton 1989: 25). Many of these changes are linked to the way society has become organized in the developed countries, such as the shift towards a more informational society where the production of information rather than goods is at a premium. Similarly, there has also been a shift towards relatively decentralized authority systems, particularly in the family, where individual members are more likely to make autonomous decisions about (among other things) what and when they eat. Finally, during the 1980s we witnessed in the United Kingdom what has been dubbed the beginnings of 'the enterprise culture', which placed importance, particularly in ideological terms, on individual entrepreneurship and initiative rather than bureaucratic surveillance and government responsibility. Gofton argues, convincingly we believe, that patterns of food consumption (and, we should add, patterns of resistance to highly processed, 'industrialized' foods) are closely related to changes in social behaviour and organization, which place private consumer choice and individualism before community-based or class-based patterns of behaviour. The trends in consumption, implied by different market 'niches', together with underlying shifts in the way households and 'communities' are constituted around specific interests, are important components of post-industrial society.

Gofton also reviews various strands of what we might term the 'sociology of food' focusing on three major areas in which attention to food might be linked to 'post-industrial' and 'post-modernist' concerns: the semiotics of food (represented by Mary Douglas in particular); the debate initiated by Pahl and Wallace (1985) and Gershuny (1985) about changes in the way the household is constituted; and the work of Bourdieu on the relationship between upbringing, class and taste (including taste in food). Something of this approach is reflected, within British anthropology, in the work of Jack Goody (1982). The work of Pahl and Gershuny, important as it is, is principally concerned with work rather than food, but the other sociological currents certainly reflect serious attention to food consumption habits, and the role of these habits in our culture. These approaches explore the phenomenology of food consumption, the way in which food tastes and habits operate not simply as indicators of lifestyle but also as ways of conveying meaning to experience. Other contributors to this debate include Charles and Kerr, whose recent work has addressed the relationship between eating habits, gender relations and the class structure, basing their analysis on empirical evidence drawn from interviews with over 200 women in the United Kingdom between 1982 and 1983 (Charles and Kerr 1988). Charles and Kerr take as their point of departure the way in which food practices contribute to the reproduction of the social order, drawing on Bourdieu and Althusser in stressing the role of ideology in social reproduction (Charles and Kerr 1988: 10). Their analysis places considerable emphasis on the meanings carried by food about the nature of the social occasion, and the way these meanings provide the ground rules for social structure, gender roles and class identities. This kind of research is phenomenological in outlook but essentially empirical, even occasionally empiricist, in design. It is grounded less in what Giddens has termed the 'knowledgeability of [the] agents' themselves (Giddens 1984: xx) and more in the unconscious behaviour of these agents, whose cultural patterns help to define them within their society.

Research such as that undertaken by Charles and Kerr delineates some important aspects of social behaviour in Britain today, but we would argue that it fails to relate behaviour within the family to wider structural changes in the role of food within society, and to the expansion of the food system. Reading Charles and Kerr (1988) one could be excused for thinking that working-class people never ate 'fast foods' (there is no entry on fast foods in the index) while 'convenience foods' are relegated to just three pages of text. Apparently the women interviewed used convenience foods, but not in the main meal they prepared – convenience foods were not 'proper meals'. A typical quotation reads:

> Expending time and effort on the production of a meal in some way conferred goodness on it: perhaps a moral rather than a nutritional goodness but this distinction

was not made by the women. . . A meal consisting of
convenience foods . . . cannot constitute a proper meal,
is by definition less healthy than a proper meal and
involves much less preparation time . . . We may
therefore see women's concern that the family should eat
properly and healthily in terms of home cooked food provided at regular intervals as having been handed down
through family tradition; it is a central part of family life.
(Charles and Kerr 1988: 130-3)

This kind of analysis places evidence of continuity before
evidence of change, and imbues the meanings people attach
to their behaviour with more importance than the behaviour
itself. It is also instructive to note how the authors have
renegotiated these meanings in the text. It is not easy to
establish from their work whether women are making more
use of convenience foods, although the suggestion is that they
probably are, because the analysis is largely concerned to
demonstrate that any resort to convenience foods needs to be
located within traditional family behaviour. Emphasizing that
people look upon food as 'natural' should not blind us to the
fact that in practice it frequently is highly processed. What
people *feel about food* may be important, but so is the evidence
about *what they eat*. One suspects that if the same women were
talking to researchers less concerned to establish its meaning
for their argument about gender and class relationships, it
would have been much more difficult to discuss food
preferences in terms of the provision of a 'proper meal'. Instead
of evidence for the reproduction of the existing social order,
we would see evidence of change, in the social changes
surrounding the way food is consumed as well as in the food
itself. Phenomenological approaches to the meaning of food
consumption would benefit from taking some cognizance of
the processes through which food itself has become an
industrally produced good.

This book has examined the effects of the modern food
system in accelerating social changes, in both developed and
developing societies. We have argued that the ideological
aspects of the food system are not confined to those usually
reported on by sociologists: the 'appropriateness' of food, and
the meaning attached to food in everyday life, within the

experience of working-class people. It is our contention that changes in the material basis of food production and consumption are linked to the transformation of 'nature', and that the establishment of new areas of commodity production influences the values attached to social relations. Capitalism is an evolving system, whose values and practices do not lie embedded in the past, anymore than the behaviour of working-class people is lodged in static definitions of what is appropriate, definitions formed by their parents and grandparents within a quite different social milieu. Food has become fully commoditized; it is therefore to be expected that its 'naturalness' would be emphasized by the food industry, in the process generating new symbolic meanings as a path to further accumulation.

As Massey (1989: ix) has observed there needs to be a debate about the 'natural', which can draw on recent post-structuralist thinking. The 'denaturalization' of space, in the form of attention to 'location theory', was a preoccupation of some social scientists, particularly geographers (Peet 1989: 43) in the 1960s. What is required now is similar attention to the way in which naturalization accompanies changes in technology and social relations. Nature has been refashioned in several ways, which this book has drawn attention to within an integrated framework of analysis.

(a) Within the labour process the employment of women has, through routinization, become naturalized as women's labour has shifted into and out of the home.
(b) Technological changes in agri-food production, now enhanced by biotechnology and genetic engineering, have manipulated biological processes, appropriating some and substituting others, redefining and refashioning nature as a source of profit and capital accumulation. As the food we consume has become more processed it has been presented as more 'natural' by the food industry.
(c) Finally, the development of modern agriculture has forced a separation between the 'countryside' as an area of consumption (amenity, recreation, aesthetics) and the production of food and fibre from the land. The drive to protect the countryside, or more generally to seek greater

protection for conservation and wilderness areas in other industrial societies, is defended and advanced on the basis of conserving what is 'natural', and a true embodiment or representation of 'nature'.

It is clear from this analysis that as the modern food system becomes more integrated, and the management of contradictions, such as those presented by the environment, become more urgent, the more the food system departs – in all three senses referred to above – from being 'natural'. Naturalization is the price society pays for seeking to transform its image of itself. Or, as Bourdieu has expressed it, each system produces the naturalization of its own arbitrariness.

THE MODERN FOOD SYSTEM WITHIN AN INTERNATIONAL PERSPECTIVE

We have argued that many of the contradictions posed by the modern food system can be handled by policy interventions, such as those to contain environmental costs and to address public concern over the quality of industrially produced food. The modern food system developed out of a combination of factors. These included: state intervention in agricultural policy designed to stimulate the adoption of new technology; guaranteed markets; the food processing industries' access to female labour, partially released from full-time employment in the home by (initially) wartime restructuring; and changes in family farming which served to legitimate the demands of more 'efficient', small-scale but highly capitalized producers. The management of the contradictions of the modern food system can be handled in the industrialized countries of the North, at a price, for several reasons that require close attention.

As we have seen, especially in Chapters 3 and 4, productivity increases in agriculture have been subsidized through industrial accumulation in the countries of the North. In the post-war period of intensive accumulation, the significance of cheap food arises not simply, as initially understood from its role, as a wage good; but also from its contribution to a widening of the market for consumer goods in general, which has accompanied the growth of labour productivity within the advanced capitalist societies. The costs of this policy, of

251

continuing to support agricultural modernization, have been passed on to the consumer and taxpayer in the European Community, the United States and Japan. It is increasingly clear that this agricultural policy model, even when only partially understood, is being called into question, notably by conservative critics as well as those with 'Greener' credentials. At the level of ideological legitimation, which has remained important throughout the post-war period in helping to define the relationship between farmers and the state, the resistance of public opinion to the priority attached to this model, at great environmental and social cost, has become an increasingly important part of the new political agenda of the 1990s.

Furthermore, the rise to importance of what we have termed 'environmental managerialism', in the industrialized countries, represents a significant policy opening, which could not have been anticipated a decade ago. The improved management of 'externalities' associated with the food system can reduce the full impact of the transformation of the countryside, through negotiated accommodations to environmental concerns. In the United Kingdom we have been conditioned to accept a managed countryside for a least a century; in other countries of the European Community the emphasis on preserving the cultural association between family farmers and the environment is equally strong politically. Indeed this nexus of interests recently gained prominence as the philosophical foundation of the European Commission's proposed rural development policy, as outlined in *The Future of Rural Society* (EC 1988). In North America, the issues are rather different, in that the separation of wildlands (to which public access is considered important) and agriculture, is more pronounced. Nevertheless, here too the armoury of interventionist policies designed to manage the contradictions of development is receiving more attention: low input farming, wilderness retention, the establishment of recreational priorities for an affluent, urban population.

Finally, as we have seen, in the industrialized countries of the North there has been considerable resistance to the complete industrialization of the diet. For a substantial minority of the population – in the United States usually considered to be about a third of the market – 'healthy' eating has become an alternative to the excesses of the 'fordist' diet. The food industry has accommodated, indeed sought to exploit, many of these changes, whose origin we see in the wider educational opportunities of the post-war generation, as well as greater private affluence. As we have seen, it is possible to increase value added in food products without necessarily losing the nutritional value of food, although this poses real problems for the large food conglomerates in increasingly competitive, oligopolistic markets (Duke 1989). Within some sectors of the durable consumer goods industries the development of 'healthy' convenience foods can also provide a stimulus to production, microwaves and toasted sandwich-makers replace or supplement other white goods, but without leading to the consumption of more unhealthy food. Changes in diet have also been prompted by wider changes in public attitudes: the increases in vegetarianism, the concern with animal welfare, the widespread interest in high-fibre, low-calorie foods as part of attempts to lose weight. These changes are significant, and are most unlikely to be reversed: they have helped to set in motion a series of effects on the supply side of the food system, which are reverberating and will continue to reverberate during the remaining years of this century and into the next.

Within the developed countries of the North, then, the modern food system poses problems for the economy and society which call for changes in policy, and the management of the environment, but which can be successfully addressed, albeit at a price in both financial and political terms. The implications of the food system for the South, we would contend, are altogether more radical. It is worth considering the differences carefully.

In the South the modern food system can only be partially adopted, as we saw in Chapter 4, because structural inequalities have prevented the industrial sector from acting as the 'motor' for the restructuring of agriculture around industrially produced food for mass consumption. In most developing countries the agricultural sector, and particularly staple production, has been appallingly neglected, not least in terms of the impact of agricultural intensification on the poor and the environment. Where rising agricultural productivity is supported by state subsidies, or subsidies from the developed countries and their institutions, it has benefited some groups much more than others. The 'fordist' diet, exemplified by the penetration of the grains–livestock complex, has only reached a very narrow band of food consumers in even the so-called 'middle-income' countries of Latin America. In most of Africa and Asia, the fordist diet remains part of the mystique of Western cultural values, as distant from its audience as the movie star.

Environmental managerialism, too, does not represent a workable strategy for most governments in the South, where it has neither tradition nor institutional foundation. In the North, the improved management of the environment, usually at the behest of relatively affluent, urban groups, is endorsed with relatively little political dissent; the modest interventions to ensure a better management of the countryside, or conservation of wilderness, have overwhelming public support in Europe and North America. In the South, on the other hand, land degradation is intimately linked to poverty in rural areas. A vast, largely rural population of poor people, many of them *very* poor, looks to the environment as the basis of its precarious livelihood. Environmental problems in developing countries can only be realistically understood as the outcome of underdevelopment and poverty. The 'environment' lacks ideological resonance; it is to be 'tamed' or 'conquered', and made to yield economic revenues and livelihoods. Environmental managerialism, without its ideological underpinnings which explain its strong cultural appeal in the North, would need to be radically modified to make any impact on the causes of environmental degradation in the South.

It can be argued, indeed, that attempts to manage the environment in developing countries, in as much as they are predicated on Northern experience, are not merely unworkable, they are also frequently prejudicial to the interests of the poor. In the Brazilian Amazon the disputed land and natural resources cannot be 'mediated' successfully by international agencies alone, whose principal concern is thought by many Brazilians (on both left and right of the political spectrum) to be the protection of the quality of life in the North, rather than the rights to livelihood of poor people in the South. Many Brazilians say, and their views are at least partly convincing, that campaigns to protect the Amazon rain forest are the new face of 'neo-imperialism', prompted by fears of global warming (largely exaggerated in view of the much more serious effect of fossil fuel emissions) and the perceived need to preserve rare animal and plant species, rather than concern about poor frontier settlers or the pernicious social impacts of the Brazilian debt.

In many parts of the South, 'resistance' is to be understood primarily in terms of the right to livelihood, to citizenship and to the recognition of basic human rights. Struggles such as those surrounding land invasions, or the protection of indigenous people and rubber tappers in the forests, represent struggles for existence, rather than movements to secure improvements in the quality of life. The development of the modern food system is linked to these struggles, albeit sometimes in tenuous ways, but the management of the environment cannot resolve the contradictions implied by the development process. Indeed, the effects of recent structural adjustment policy, and attempts to manage the enormity of the debt burden in Latin America and Africa in particular, have intensified environmental problems, and make any 'solution' that does not include major changes in these policies largely irrelevant. The implications of the partial adoption of elements of the modern food system in the South can only be understood, and the resulting problems successfully addressed, by changes in structural policies encompassing the global economy. They cannot be resolved, as to a large extent they can in the North, where they occur.

Finally, as far as changes in the diet are concerned, the partial development of the modern food system in the countries of the South has served to replace (and in some cases destroy) existing, indigenous food systems. This has occurred without effective resistance from a broadly based mass movement of consumers, educated as to their rights, and able to exert pressure on government to accede to their demands. It is virtually impossible to envisage public campaigns over salmonella or food additives in Mexico or Indonesia, or, what is altogether more telling, effective mass movements to introduce new standards of food hygiene in shops or on street corners. Meanwhile, food irradiation – increasingly contested by the radical food lobby in the North – opens up market opportunities in the South.

It is clear, then, that what is at issue if we consider North–South relations, and examine the implications of the development of the modern food system within an international framework, is a set of concerns which stretch the parameters of our model, and our analysis. To speak of resistance and struggle in the South, is to speak of the survival of alternative systems which link the production of food directly to its transformation and consumption, systems which embody quite different patterns of social reproduction, systems which – in the form of ubiquitous 'externalities' – carry many of the environmental and social costs of our 'successful' development model. Our ability to grasp the fact that the economic and political hegemony of our values, and our economic policies, have served to undermine the integrity of other food systems, remains the major challenge we in the North face today. For the spread of the modern food system marks the success of the North's cultural hegemony, as well as its economic supremacy. We need to address the sustainability of our own models, then, before we are in a position to hand them down to others.

BIBLIOGRAPHY

Aglietta, M. (1979) *A Theory of Capitalist Regulation*, London: New Left Books.

Agristats (1984) *Agristats*, Wye, Kent: Wye College (University of London).

Allen, G.E. (1978) *Life Science in the Twentieth Century*, Cambridge: Cambridge University Press.

—— (1979) 'The transformation of a science: T.H. Morgan and the emergence of a new American biology', in A. Oleson and J. Koss (eds) *The Organization of Knowledge in Modern America, 1860–1920*, Baltimore: Johns Hopkins University Press.

Anderson, P. (1987) 'The figures of descent', *New Left Review*, January/February.

Baker, N. (1988) *The Mezzanine*, Cambridge: Granta.

Baldock, D. (1984) *Wetland Drainage in Europe*, London: Institute of European Environmental Policy.

Barbier, E. (1989) *Economics, Natural Resource Scarcity and Development*, London: Earthscan/International Institute for Environment and Development.

Barrington Moore, W. (1969) *Social Origins of Dictatorship in Democracy*, Harmondsworth: Penguin.

Barton, J.H. (1989) 'Legal trends and agricultural biotechnology: effects on developing countries', *Trends in Biotechnology* 7, October.

Baxter, S. and Raw, G. (1988) 'Fast food, fettered work: Chinese women in the ethnic catering industry', in S. Westwood and P. Bhachu (eds) *Enterprising Women*, London: Routledge.

Belasco, W.J. (1989) *Appetite for Change*, New York: Pantheon Books.

Beneria, L. and Stimpson, C. (1987) (eds) *Women, Households and the Economy*, New Brunswick: Rutgers University Press.

Benoist, J.R. de (1983) 'Manger sénégalais', *Croissance des Jeunes Nations*, no. 254, October.

Berlan, J.-P. (1991) 'The historical roots of the present agricultural crisis', in L. Busch, F.H. Buttel, W.H. Friedland and A. Rudy (eds) *The New Political Economy of Advanced Capitalist Agriculture*, Boulder, Colo.: Westview Press.

257

BIBLIOGRAPHY

Berlan, J.-P. and Lewontin, R.C. (1986) 'Breeders rights and patenting life forms', *Nature* 332 (6082), 28 August.

Berlan, J.-P., Bertrand, J.-P. and Lebas, L. (1977) 'The growth of the American soybean complex', *European Review of Agricultural Economics* 4 (4).

Bertrand, J.-P. (1985) 'Brésil: modernisation agricole et restructuration alimentaire dans la crise internationale', *Revue Tiers Monde* 26 (104).

Biggs, S. (1988) 'Research and development: on the farm and in the agricultural research centre', paper delivered at Wye College (University of London), unpublished ms.

Bishop, J. and Allen, J. (1989) 'The on-site costs of soil erosion in Mali: a natural resource accounting approach', *Environmental Working Paper*, Washington, DC: World Bank.

Blaikie, P. (1985) *The Political Economy of Soil Erosion in Developing Countries*, London: Longman.

BNDES (1988) *Mudanças Estruturais Nas Atividades Agrarias: Uma Analise das Relações Intersetoriais no Complexo Agroindustrial Brasileiro*, *Estudos BNDES* 9, Rio de Janeiro: Banco National de Desenvolvimento Economica e Social.

Bogue, A.G. (1983) 'Changes in mechanical and plant technology: the corn belt, 1910–40', *The Journal of Economic History* 43 (1).

Bonnifield, P. (1979) *The Dust Bowl: Men, Dirt and Depression*, Albuquerque: University of New Mexico Press.

Bose, C. (1982) 'Technology and changes in the division of labour in the American home', in E. Whitelegg *et al.* (eds) *The Changing Experience of Women*, Oxford: Martin Robertson/The Open University.

Bowers, J.K. (1985) 'British agricultural policy since the Second World War', *The Agricultural History Review*, vol. 33, part 1.

Bowers, J.K. and Cheshire, P. (1983) *Agriculture, the Countryside and Land Use*, London: Methuen.

Bradley, A. (1984) 'Segmentation in local labour markets', in T. Bradley and P. Lowe (eds) *Locality and Rurality: Economy and Society in Rural Regions*, Norwich: Geo Books.

—— (1985) 'Poverty and dependency in village England', in P. Lowe, T. Bradley and S. Wright (eds) *Deprivation and Welfare in Rural Areas*, Norwich: Geo Books.

Brady, N. (1989) 'Making tropical agriculture sustainable: the CIAT experience', Washington DC: World Resources Institute.

Bray, J.O. and Watkins, P. (1964) 'Technical change in corn production in the United States, 1870–1960', *Journal of Farm Economics* 46, November.

Briggs, A. (1988) *Victorian Things*, London: Batsford.

Brouwer, R. and Jansen, K. (1988) *Simulatie van Participatie: Farming Systems Research*, SG-Paper 103, Studium Generale, Landbouwuniversiteit Wageningen.

Bull, A.T., Holt, G. and Lilley, M.D. (1982) *Biotechnology: International Trends and Perspectives*, Paris, Organization for Economic Co-operation and Development.

Bunker, S. (1985) *Underdeveloping the Amazon*, Chicago: Chicago University Press.

Burch, W.B. (1983) 'Time, habitat and social structure', in T. O' Riordan and K. Turner (eds) *An Annotated Reader in Environmental Planning and Management*, Oxford, Pergamon.

Burnett, J. (1989) *Plenty and Want: A Social History of Food in England from 1815 to the Present Day* (2nd edn), London: Routledge.

Burns, J. (1983) 'A synoptic view of the food industry', in J. Burns, J. McInerney and A. Swinbank (eds) *The Food Industry*, London: Heinemann.

Burrill, G.S. (1989) 'Biotechnology – a worldwide perspective', *Biotech 89 Proceedings*, Pinner, Middlesex: Blenheim Online.

Busch, L. and Lacy, W.B. (eds) (1986) *The Agricultural Scientific Enterprise: A System in Transition*, Boulder, Colo.: Westview Press.

Busch, L., Bonnano, A. and Lacy, W. (1989) 'Science, technology, and the restructuring of agriculture', *Sociologia Ruralis* 29 (2).

Buttel, F.H. (1982) *Environmental Quality in Agriculture: Some Observations on Political-Economic Constraints on Sustainable Resource Management*, Bulletin no. 128, Ithaca, NY: Cornell University Department of Rural Sociology.

—— (1986a) 'Biotechnology and agricultural research policy: emergent issues; in K.A. Dahlberg (ed.) *New Directions for Agriculture and Agricultural Research*, Totowa, NJ: Rowman & Allenheld.

—— (1986b) 'Biotechnology and agricultural research policy', in V.J. Rhodes (ed.) *Agricultural Science Policy in Transition*, Bethesda, MD: Agricultural Research Institute.

—— (1989a) 'The US farm crisis and the restructuring of American agriculture: domestic and international dimensions', in D. E. Goodman and M. R. Redclift (eds) *The International Farm Crisis*, London: Macmillan.

—— (1989b) *Rethinking Biotechnology Policy*, unpublished paper.

Buttel, F. H. and Busch, L. (1988) 'The public research system at the crossroads', *Agricultural History*, Spring.

Caballero, J.M. (1984) 'Agriculture and the peasantry under industrialisation pressures: lessons from the Peruvian experience', *Latin American Research Review* 19 (2).

Capra, F. (1976) *The Tao of Physics*, London: Collins.

Carter, D. (1989) 'Developments in grocery retailing', *British Food Journal* 91 (1): 3–15.

Chambers, R. and Jiggins, J. (1987) 'Agricultural research for resource-poor farmers; Part 1: Transfer-of-technology and farming systems research', *Agricultural Administration and Extension* 27.

Charles, N. and Kerr, M. (1988) *Women, Food and Families*, Manchester: Manchester University Press.

Charvet, J.-P. (1987) *Le Désordre Alimentaire Mondial*, Paris: Hatier.

Cherfas, J. (1982) *Man Made Life: A Genetic Engineering Primer*, Oxford: Basil Blakwell.

Cheshire, P. (1985) 'The environmental implications of European agricultural support policies', in D. Baldock and D. Conder (eds) *Does the CAP fit the Environment?*, London: Institute for European Environmental Policy.

Clairemonte, F.F. and Cavanagh, J.J. (1982) 'Transnational corporations and global markets: changing power relations', *Trade and Development* 4, Winter.

Cochrane, W.W. (1979) *The Development of American Agriculture: A Historical Analysis*, Minneapolis: University of Minnesota Press.

Cochrane, W.W. and Ryan, M. E. (1981) *American Farm Policy, 1948–1973*, Minneapolis: University of Minnesota Press.

Cockburn, C. (1985) *Machinery of Dominance*, London: Pluto Press.

Collins, E. (1976) 'The *consumer revolution* and the growth of factory foods: changing patterns of bread and cereal eating in Britain in the twentieth century', in D. Oddy and D. Miller (eds) *The Making of the Modern British Diet*, London: Croom Helm.

Connor, S. (1988) 'The battle for Britain's biotechnology', *New Scientist* 119 (1625), 11 August.

Conran, S. (1977) *Superwoman*, Harmondsworth: Penguin.

Conway, G.R. and Barbier, E.B. (1988) 'After the Green Revolution: sustainable and equitable agricultural development', *Futures* 20 (6).

Cook, A.G. (1989) 'Patents as non-tariff trade barriers', *Trends in Biotechnology* 7, October.

Cornia, G.A., Jolly, R. and Stewart, F. (1987) *Adjustment with a Human Face*, vol. 1, Oxford: Clarendon Press.

Corley, T. (1976) 'Nutrition, technology and the growth of the British biscuit industry, 1820–1900', in D. Oddy and D. Miller (eds) *The Making of the Modern British Diet*, London: Croom Helm.

Cornuel, D. and Duriez, B. (1985) 'Local exchange and state intervention', in N. Redclift and E. Mingione (eds) *Beyond Employment*, Oxford: Blackwell.

Countryside Commission (1987) National Countryside Recreation Survey, *Social Trends* 17.

Covey, D. (1987) 'Employment in the food chain', paper presented to Rural Economy and Society Study Group, Annual Conference, Loughborough.

Cox, G., Lowe, P. and Winter, M. (1989) 'The farm crisis in Britain', in D. Goodman and M. R. Redclift (eds) *The International Farm Crisis*, London: Macmillan.

CRS (1984) *Recent Advances in the Plant Sciences: Applications to Agriculture and Agricultural Productions*, Congressional Research Service, a report for the Committee on Science and Technology, US House of Representatives, Washington, DC: US Government Printing Office.

Dade, P. (1988) 'Interesting times – the new consumer, curse or opportunity?', *British Food Journal* 90 (3).

Darling, G. (1941) *The Politics of Food*, London: Labour Book Service.

Daly, L. and Beharrell, B. (1988) 'Health , diet and the marketing of food and drink – some theoretical problems', *British Food Journal* 90 (1): 5–9.

Dawes, F.V. (1984) *Not in Front of the Servants: A True Portrait of Upstairs, Downstairs Life*, London: Hutchinson.

Dawson, J.A., Shaw, S.A. and Rana, J. (1988) 'Future trends in food retailing: results of a survey of retailers', *British Food Journal* 91 (2): 51–7.

Day, R.H. (1967) 'The economics of technological change and the demise of the share-cropper', *The American Economic Review* 57 (3), June.

de Janvry, A. (1981) *The Agrarian Question and Reformism in Latin America*, Baltimore: Johns Hopkins University Press.

de Janvry, A. and Sadoulet, E. (1988) 'Investment strategies to combat rural poverty: a proposal for Latin America', paper presented at the Seventh World Congress for Rural Sociology, Bologna.

de Janvry, A., Runsten, D. and Sadoulet, E. (1987) *Technological Innovations in Latin American Agriculture*, San José, Costa Rica: Inter-American Institute for Cooperation on Agriculture.

Deo, S.D. and Swanson, L.E. (1987) *The Political Economy of Agricultural Research in the Third World*, paper presented at the annual meetings of the Rural Sociological Society, Madison, Wisconsin.

Destler, I.M. (1978) 'United States food policy 1972–76: reconciling domestic and international objectives', *International Organization* 32 (3), Summer.

Devall, B. and Sessions, W. (1985) *Deep Ecology*, Layton, Utah: Peregrine Smith.

de Vroey, M. (1984) 'A regulation approach interpretation of contemporary crisis', *Capital and Class*, no. 23, Summer.

Dibner, M. (1989) 'Biotechnology in the United States and Japan: strategies and trends', *Biotech 89 Proceedings*, Pinner, Middlesex: Blenheim Online.

Duke, R.C. (1989) 'A structural analysis of the UK grocery retail market', *British Food Journal* 91 (5): 17–22.

Dumont, R. (1970) *Types of Rural Economy: Studies in World Agriculture*, London: Methuen.

ENDS (1988) 'Biotechnology 6: Scientists disagree over the environmental hazards of genetic engineering', *Report 159*, April.

Engels, F. (1892) *The Condition of the Working Class in England in 1844*, London: George Allen & Unwin.

Esquinas-Alcazar, J. (1989) *FAO Global System on Plant Genetic Resources*, paper presented at the CTA/FAO Symposium on Plant Biotechnologies for Developing Countries, Luxembourg.

ESRC (1989) *The Countryside in Question: A Research Strategy*, Economic and Social Research Council, Countryside Change Initiative, Working Paper no. 1, November.

Etzexarreta, M. and Viladomiu, L. (1989) 'The restructuring of Spanish agriculture and Spain's accession to the EEC', in D. Goodman and M.R. Redclift (eds) *The International Farm Crisis*, London: Macmillan.

European Commission (1985) 'Council Regulation 797/85: on improving the efficiency of agricultural structures', *Official Journal of the EEC*, L93/1–13, Brussels: European Commission.

—— (1986) *A Future for Community Agriculture: Commission Guidelines*, Brussels: European Commission.

European Community (1985–6) *Report on behalf of the Committee on the Environment, Public Health and Consumer Protection*, European Parliament Working Document (A-2-207/85), Brussels: European Community.

FAO (1988) *Integration of Environmental Aspects in Agricultural, Forestry and Fishery Policies in Europe*, Food and Agriculture Organization, XVI Regional Conference for Europe, Cracow, Poland, 23–26 August.

—— (1989a) *Implications of New Biotechnologies for the International Undertaking*, Commission on Plant Genetic Resources, Third Session, 17–21 April, CPGR/89.9, Rome: Food and Agriculture Organization.

—— (1989b) *Report of the Commission on Plant Genetic Resources*, Third Session, 17–21 April, CPGR/89/Rep., Rome: Food and Agriculture Organization.

Farrell, K.R. and Runge, C.F. (1983) 'Institutional innovation and technical change in American agriculture: the role of the New Deal', *American Journal of Agricultural Economics*, December.

Feyerabend, P. (1987) *Farewell to Reason*, London: Verso.

Finegold, K. (1985) *Agriculture, State, Party, and Economic Crisis: American Farm Policy and the Great Depression*, unpublished ms.

Fite, G.C. (1980) 'Mechanization of cotton production since World War II', *Journal of Agricultural History* 54 (1), January.

—— (1981) *American Farmers: The New Minority*, Bloomington: Indiana University Press.

Fowler, C., Lachkovics, E., Mooney, P.R. and Shand, H. (1988) *The Laws of Life: Another Development and the New Biotechnologies*, special issue, *Development Dialogue* 1–2.

Frank, J. and Wheelock, V. (1988) 'International trends in food consumption', *British Food Journal* 90 (1): 22–9.

Frank, W. (1983) 'Part-time farming, underemployment and double activity in the EEC', *Sociologia Ruralis* xxiii.

Franklin, S.H. (1970) *The European Peasantry: The Final Phase*, London: Methuen.

Friedmann, H. (1982) 'The political economy of food: the rise and fall of the postwar international food order', *American Journal of Sociology* 88: 246–86, supplement.

—— (1987) 'Everyone's food comes from abroad: class-specific diets and the international specialization of agricultural production', paper presented at the Rural Sociology Society annual meetings, Madison, Wisconsin.

Friedmann, H. and McMichael P. (1989) 'Agriculture and the state system: rise and decline of national agricultures, 1870 to the present', *Sociologia Ruralis* xxix (2).

Galeski, B. (1972) *Basic Concepts of Rural Sociology*, Manchester: Manchester University Press.

Gardiner, J., Himmelweit, S. and Mackintosh, M. (1980) 'Women's domestic labour', in E. Malos (ed) *The Politics of Housework*, London and New York: Allison & Busby.

262

Gershuny, J.I. (1985) 'Economic development and change in the mode of provision of services', in N. Redclift and E. Mingione (eds) *Beyond Employment*, Oxford: Basil Blackwell.

Giddens, A. (1976) *The Class Structure of the Advanced Societies*, Cambridge: Cambridge Universtiy Press.

—— (1984) *The Constitution of Society*, Oxford: Polity Press.

Glyn, A., Hughes, A., Lipietz, A. and Singh, A. (1988) *The Rise and Fall of the Golden Age*, Department of Applied Economics, Working Paper no. 884, University of Chicago.

Gofton, L. (1989) 'Sociology and food consumption', *British Food Journal* 91 (1).

Goldsmith, E. and Hildyard, N. (1986) *Green Britain or Industrial Wasteland*, London: Polity Press.

Goliber, T.J. (1989) 'Africa's expanding population: old problems, new policies', *Population Bulletin* 44 (3), November.

Goodman, D.E. and Hall, A.L. (1990) *The Future of Amazonia: Destruction or Sustainable Development?*, London: Macmillan.

Goodman, D.E. and Redclift, M.R. (1981) *From Peasant to Proletarian: Capitalist Development and Agrarian Transitions*, Oxford: Basil Blackwell.

—— (1985) 'Capitalism, petty commodity production and the farm enterprise', in G. Cox, P. Lowe and M. Winter (eds) *Agriculture, People and Policies*, London: Allen & Unwin.

—— (eds) (1989) *The International Farm Crisis*, London: Macmillan.

Goodman, D.E. and Wilkinson, J. (1990) 'Patterns of research and innovation in the modern agri-food system', in P. Lowe, T.K. Marsden and S. Whatmore (eds) *Technological Change and the Rural Environment*, London: David Fulton.

Goodman, D.E., Sorj, B. and Wilkinson, J. (1987) *From Farming to Biotechnology: A Theory of Agro-industrial Development*, Oxford: Basil Blackwell.

Goody, J. (1982) *Cooking, Cuisine and Class*, Cambridge: Cambridge University Press.

Goonatilake, S. (1984) *Aborted Discovery*, London: Zed Press.

Gorz, A. (1989) 'Ecology and environmentalism', *The New Statesman*, 2 June.

Grigg, D. (1986) *The World Food Problem, 1950–80*, Oxford: Basil Blackwell.

Griliches, Z. (1957) 'Hybrid corn: an exploration in the economics of technological change', *Econometrica* 25 (4), October.

Guha, R. (1989) *The Unquiet Woods: Ecological Change and Peasant Resistance in the Himalayas*, Delhi: Oxford University Press.

Hadwiger, D.F. (1970) *Federal Wheat Commodity Programs*, Ames, Iowa: Iowa Universtiy Press.

Hall, A.L. (1989) *Developing Amazonia*, Manchester: Manchester University Press.

Halsey, A.H. (1987) 'Social trends since World War II', *Social Trends* 17: 11–19.

Hamilton, P. (1985) 'Small farmers and food production in Western Europe', *International Social Science Journal* xxxvii (3): 345–60.

Hamnett, C., McDowall, L. and Sarre, P. (eds) (1989) *The Changing Social Structure*, London: Sage/The Open University.
Hartmann, H. (1974) 'Capitalism and women's work in the home, 1900–1930', Ph.D. dissertation, Yale University.
Hayami, Y. and Ruttan, V. W. (1971) *Agricultural Development: An International Perspective*, Baltimore: Johns Hopkins University Press.
Heady, E.O. (1982) 'The adequacy of agricultural land: a demand-supply perspective', in P.R. Crosson (ed.) *The Cropland Crisis: Myth or Reality?*, Baltimore: Resources for the Future, Inc.
Healey, M.J. and Ilbery, B.W. (eds) *The Industrialization of the Countryside*, Norwich: Geo Books.
Heasman, M. (1989) 'Sugar and the modern food system', *British Food Journal* 91 (3): 9–16.
Hecht, S. and Cockburn, A. (1989) *The Fate of the Forest: Developers, Destroyers and Defenders of the Amazon*, London: Verso.
Hobsbawm, E. (1979) *The Age of Capital, 1848–1875*, Harmondsworth: Penguin.
Homem de Melo, F. (1983) *O Problema Alimentar No Brasil*, Rio de Janeiro: Paz e Terra.
Hooper, M. (1984) 'What are the main impacts of agriculture on wildlife?', in D. Jenkins (ed.) *Agriculture and the Environment*, Cambridge Institute of Terrestrial Ecology.
Hopkins, R.F. and Puchala, D.J. (1978) 'Perspectives on the international relations of food', in R.F. Hopkins and D.J. Puchala (eds) *The Global Political Economy of Food*, Madison: University of Wisconsin Press.
Howes, M. (1979) 'The use of indigenous technical knowledge in development', *Institute of Development Studies Bulletin* 10 (2).
Hughes, J. D. (1975) *Ecology and Ancient Civilizations*, Albuquerque: University of New Mexico Press.
Huws, U. (1985) 'Challenging commoditisation', in Collective Design (eds) *Very Nice Work If You Can Get It*, Nottingham: Spokesman.
IAMM (1987) *Les Cents Premiers Groupes Agro-industriels Mondiaux*, Montpellier: Institut Agronomique Mediterranéen de Montpellier.
IBRD (1979) *Capital Markets Study*, Washington, DC: International Bank for Reconstruction and Development, mimeo.
—— (1986) *World Development Report 1986*, New York: Oxford University Press.
IDB (1986) *Economic and Social Progress in Latin America, 1986 Report* Washington DC: Inter-American Development Bank.
IFPRI (1981) International Food Policy Research Institute, *IFPRI Report* 9 (3), October.
Illich, I. (1975) *The Limits of Medicine*, Harmondsworth: Penguin.
International Wheat Council (1988) *Long-term Outlook for Grain Imports by Developing Countries*, Secretariat Paper no. 17: London: International Wheat Council.
Jefferies, R. (1987) *Round About A Great Estate*, Bradford on Avon: Ex Libris Press.
Jenkins, D. (1988) 'The kitchen of the future', *British Food Journal* 90 (6): 255–8.

Johnston, B.F. and Kilby, P. (1975) *Agriculture and Structural Transformation*, London: Oxford University Press.

Johnston, J. (1977) *A Hundred Years of Eating*, Dublin: Gill & Macmillan.

Johnstone, G.N. (1976) 'The growth of the sugar trade and refining industry', in D. Oddy and D. Miller (eds) *The Making of the Modern British Diet*, London: Croom Helm.

Juma, C. (1989) *The Gene Hunters*, London: Zed Books.

Kay, G. (1975) *Development and Underdevelopment: A Marxist Analysis*, London: Macmillan.

Kenney, M. (1989) *Biotechnology: The University–Industrial Complex*, New Haven, Conn.: Yale University Press.

Kenney, M., Kloppenburg, J. Jr, Buttel, F.H. and Cowan, J.T. (1983) 'Genetic engineering and agriculture', *Biotech 83 Proceedings*, London: Online Publications.

Kenney, M., Lobao, L.M., Curry, K. and Goe, W.R. (1989) 'Midwestern agriculture in US fordism: from the New Deal to economic restructuring', *Sociologia Ruralis* 29 (2).

Kidd, G. (1987) 'Macroeconomics of seed and plant biotechnology in the 1990s', *Biotech 87 Proceedings*, London: Online Publications.

Kindleberger, C.P. (1951) 'Group behaviour and international trade', *Journal of Political Economy* 59 (1).

—— (1987) *The World in Depression, 1929–39*, Harmondsworth: Penguin.

Kingman, S. (1989) 'Plant patent faces new legal challenge', *New Scientist* 124 (1695), 16 December.

Klausner, A. (1986) 'MONSANTO: betting a giant on biotechnology', *Bio/Technology*, May.

Kloppenburg, J. Jr (1984) 'The social impacts of biogenetic technology in agriculture: past and future ', in G.M. Berardi and C.C. Geisler (eds) *The Social Consequences and Challenges of New Agricultural Technologies*, Boulder, Colo.: Westview Press.

—— (1988a) *First the Seed: The Political Economy of Plant Biotechnology*, Cambridge: Cambridge University Press.

—— (ed.) (1988b) *Seeds and Sovereignty: The Use and Control of Plant Genetic Resources*, Durham and London: Duke University Press.

Lajo, M. (1978) *Industria agroalimentaria y transnacionales: el caso peruano*, Lima: Pontificia Universidad Catolica.

Lang, T. and Wiggins, P. (1985) 'The industrialisation of the UK food system: from production to consumption', in M.J. Healey and B.W. Ilbery (eds) *The Industrialisation of the Countryside*, Norwich: Geo Books.

Lindert, P.H. and Williamson, J.G. (1980) *American Inequality: A Macroeconomic History*, New York: Academic Press.

Lockeretz, W. (1988) 'Agricultural diversification by crop introduction: the US experience with the soybean', *Food Policy* 13 (2), May.

Lowe, P., Bradley, T. and Wright, S. (eds) (1985) *Deprivation and Welfare in Rural Areas*, Norwich: Geo Books.

Lowe, P., Cox, G., MacEwan, M., O' Riordan, T. and Winter, M. (1986) *Countryside Conflicts: The Politics of Farming, Forestry and*

Conservation, Aldershot: Gower.
Lu, Y., Cline, P. and Quance, L. (1979) *Prospects for Productivity Growth in American Agriculture*, Washington, DC: US Department of Agriculture.
Luke, T. (1988) 'The dreams of Deep Ecology', *Telos* 76, Summer.
McDowall, L. (1989) 'Gender divisions', in C. Hamnett, L. McDowell and P. Sarne (eds) *The Changing Social Structure*, London: Sage/The Open University.
MacEwan, A. and MacEwan, M. (1982) *National Parks: Conservation or Cosmetics?*, London: Allen & Unwin.
McLaughlin, B. (1986) 'The rhetoric and reality of rural deprivation', *Journal of Rural Studies* 2 (4): 363–82.
Maddison, A (1982) *Phases of Capitalist Development*, Oxford: Oxford University Press.
MAFF (1975) *Food from our Own Resources*, Cmnd 6020, London: Her Majesty's Stationery Office.
—— (1979) *Famine and the Nation*, Cmnd 7458, London: Her Majesty's Stationary Office.
Maffei, E., Ferreira Irmão, J. and Ramos de Souza, H. (1986) *Irrigação e Emprego no Sertão do São Francisco*, Recife: OIT/PNUD/SUDENE.
Malos, E. (ed.) (1980) *The Politics of Housework*, London and New York: Allison & Busby.
Marsden, T., Lowe, P. and Munton, R. (1990) 'The social and economic restructuring of rural Britain: a position statement', ESRC Countryside Change Initiative.
Marsden, T., Munton, R. and Whatmore, S (1987) 'Uneven development and the restructuring process in British agriculture: a preliminary exploration', *Journal of Rural Studies* 3: 297–308.
Martins, S.W. (1988) *Norfolk: A Changing Countryside*, Chichester: Phillimore.
Massey, D. (1989) 'Foreword', in R. Peet and N. Thrift (eds) *New Models of Geography*, vol. I, London: Unwin Hyman.
Massey, D. and Catalano, A. (1978) *Capital and Land: Landownership by Capital in Great Britain*, London: Edward Arnold.
Mathias, P. (1976) 'The British tea trade in the nineteenth century', in D. Oddy and D. Miller (eds) *The Making of the Modern British Diet*, London: Croom Helm.
Mayo, N. and Denison-Ender, N. (1988) 'Development of, and demand for, new food and drink products', *British Food Journal* 90 (5).
Meissner, M., Humphrey, E.W., Meis, S.M. and Schen, W.J. (1988) 'No exit for wives: sexual division of labour and the cumulation of household demands in Canada', in R.E. Pahl (ed.) *On Work*, Oxford: Blackwell.
Mendras, H. (1970) *The Vanishing Peasant: Innovation and Change in French Agriculture*, Cambridge, Mass.: MIT Press.
Mingione, E. (1985) 'Social reproduction of the surplus labour force', in N. Redclift and E. Mingione (eds) *Beyond Employment*, Oxford: Blackwell.

Mond, Sir A. (1927) *Industry and Politics*, London: Macmillan.

Mooney, P.R. (1979) *Seeds of the Earth: A Private or Public Resource?*, Ottawa: Inter Pares.

—— (1983) *The Law of the Seed: Another Development and Plant Genetic Resources, Development Dialogue* 1–2, special issue.

Moseley, M. (1980) 'Rural deprivation and its relevance to the inner city debate', Inner City Papers 9, Social Science Research Council.

Moses, P.B., Tavares, J.E. and Hess, C.E. (1988) 'Funding agricultural biotechnology research', *Bio/Technology*, February.

Munton, R., Whatmore, S. and Marsden, T. (1987) 'Finance capital and the food complex', paper delivered to the Annual Conference of the Rural Economy and Society Study Group, Loughborough, December.

Murcott, A. (1983) *The Sociology of Food and Eating*, Aldershot: Gower.

Myers, N. (1981) 'The hamburger connection: how Central America's forests become North America's hamburgers', *Ambio* 10 (1).

Nash, R. (1967) *Wilderness and the American Mind*, New Haven, Conn.: Yale University Press.

Nations, J.D. and Komer, D.I. (1987) 'Rainforests and the hamburger society', *The Ecologist* 17 (4–5).

NCC (1977) *Nature Conservation and Agriculture*, London: Nature Conservancy Council.

—— (1984) *Nature Conservation in Great Britain*, London: Nature Conservancy Council.

National Consumer Council (1988) *Consumers and the Common Agricultural Policy*, National Consumer Council, London: Her Majesty's Stationery Office.

Newby, H. (1977) *The Deferential Worker*, London: Allen Lane.

—— (1983) 'Living from hand to mouth: the farmworker, food and agribusiness', in A. Murcott (ed.) *The Sociology of Food and Eating*, Aldershot: Gower.

Newby, H. and Buttel, F. (1980) 'Towards a critical rural sociology', in F.H. Buttel and H. Newby (eds) *The Rural Sociology of the Advanced Societies*, London: Croom Helm.

Newby, H., Bell, C., Rose, P. and Saunders, P. (1978) *Property, Paternalism and Power*, London: Hutchinson.

Nix, J. (1984) *Farm Management Pocketbook*, Wye, Kent: Farm Business Unit, Wye College (University of London).

Nordskog, A.W. (1977) 'Success and failure of quantitative genetic theory in poultry', in E. Pollacket *et al.* (eds) *Proceedings of the International Conference on Quantitative Genetics*, Ames, Iowa: Iowa University Press.

North, D.C. (1958) 'Ocean freight rates and economic development, 1750–1913', *Journal of Economic History* 18 (4).

O' Connor, J. (1988) 'Uneven and combined development: a view from the South', *Race and Class* 30 (3).

O' Day, R. (1985) *The Changing Experience of Women*, course units, Milton Keynes: Open University Press.

Oddy, D. (1976) 'A nutritional analysis of historical evidence: the working class diet, 1880–1914', in D. Oddy and D. Miller (eds) *The Making of the Modern British Diet*, London: Croom Helm.

Oddy, D. and Miller, D. (eds) (1976) *The Making of the Modern British Diet*, London: Croom Helm.

OECD (1989) 'Agricultural and environmental policies: opportunities for integration', Paris: Organization for Economic Co-operation and Development.

Oman, C. and Rama, R. (1988) 'The food industry', in C. Oman (ed.) *New Forms of Investment in the Mining, Petrochemicals, Automotive, Textiles and Food Industries in Developing Countries*, Paris: OECD Development Centre, mimeo.

OTA (1981) *Impacts of Applied Genetics*, Washington, DC: US Congress.

Othick, J. (1976) 'The cocoa and chocolate industry in the nineteenth century', in D. Oddy and D. Miller (eds) *The Making of the Modern British Diet*, London: Croom Helm.

Pahl, J. (1970) 'Community and locality', in *Patterns of Urban Life*, London: Longman.

—— (1983) 'The allocation of money and the structuring of inequality within marriage', *Sociological Review* 3: 237–62.

—— (1984) *Division of Labour*, Oxford: Basil Blackwell.

—— (1989) *Money and Marriage*, London: Macmillan.

Pahl, J. and Wallace, C. (1985) 'Household work strategies in economic recession', in N. Redclift and E. Mingione (eds) *Beyond Employment*, Oxford: Basil Blackwell.

Paquette, R. (1984) 'Rapidité des changements dans les habitudes de consommation alimentaire à Kingston, Jamaique', *25ᵉ Congrés International de Geographie*, Paris.

Parker, W. N. and Klein, J. L. V. (1966) 'Productivity growth in grain production in the United States, 1840–60 and 1900–1910', *Output, Employment and Productivity in the United States after 1900*, NBER Studies in Income and Wealth, vol. 30, New York: Columbia University Press.

Pearce, D., Barbier, E. and Markandaya, A. (1989) *Blueprint for a Green Economy*, London: Earthscan.

Pearse, A. (1980) *Seeds of Plenty, Seeds of Want: Social and Economic Implications of the Green Revolution*, Oxford: Clarendon Press.

Peet, R. (1989) 'Introduction', in R. Peet and N. Thrift (eds) *New Models in Geography*, London: Unwin Hyman.

Perkins, J. A. (1986) 'Dualism in German agrarian historiography', *Comparative Studies* 28 (3): 287–306.

Perrot, R. (1968) *The Aristocrats*, London: Weidenfeld & Nicolson.

Perutz, M. (1987) 'The birth of molecular biology', *New Scientist* 114 (1561), 21 May.

Platteau, J.-P. (1988) *The Food Crisis in Africa: A Comparative Structural Analysis*, WIDER Working Papers, WP44, April, Helsinki: World Institute for Development Economics Research.

Potter, C. (1989) 'Approaching limits: farming contraction and environmental conservation in the UK', in D. Goodman and

BIBLIOGRAPHY

M. Redclift (eds) *The International Farm Crisis*, London: Macmillan.
—— (1990) 'Conservation under a European farm survival policy', Wye College (University of London), unpublished ms.
Rama, R. (1985) 'Some effects of the internationalization of agriculture on the Mexican agricultural crisis', in S.E. Sanderson (ed.) *The Americas in the International Division of Labour*, New York: Holmes & Meier.
Rappaport, R. (1967) *Pigs for Ancestors*, New Haven, Conn.: Yale University Press.
Rasmussen, W. D. (1962) 'The impact of technological change in American agriculture, 1862–1962', *Journal of Economic History* 22 (4).
Ratner, M. (1989) 'Crop Biotech '89: research efforts are market driven', *Bio/Technology*, April.
Read, N. (1989) 'The "near market" concept applied to UK agricultural research', *Science and Public Policy* 16 (4): 233–8.
Redclift, M.R. (1978) *Agrarian Reform and Peasant Organisation on the Ecuadorian Coast*, London: Athlone Press.
—— (1984) *Development and the Environmental Crisis*, London: Methuen.
—— (1989) 'Turning nightmares into dreams: the Green Movement in Eastern Europe', *The Ecologist* 19 (5): 177–83.
—— (1990) 'Developing sustainability: designating agro-ecological zones', *Land Use Policy* 7 (3), July.
—— (1991) 'Making waves: sustainable development meets its critics', in F. Buttel and L. Thrupp (eds) *The Greening of Development*, Ithaca, NY: Cornell University Press.
Reeves. M. (1913) *Round About A Pound A Week*, London.
Repetto, R., Magrath, W., Wells, M., Beer, C. and Rossini, F. (1989) *Wasting Assets: Natural Resources in the National Accounts*, Washington, DC: World Resources Institute.
Revel, A. and Riboud, C. (1986) *American Green Power*, Baltimore: John Hopkins University Press.
Rice, M. S. (1939) 'Working class wives', in E. Malos (ed.) (1980) *The Politics of Housework*, London and New York: Allison & Busby.
Richardson, T.L. (1976) 'The agricultural labourer's standard of living in Kent', in D. Oddy and D. Miller (eds) *The Making of the Modern British Diet*, London: Croom Helm.
Rockefeller Foundation (1982) *Science for Agriculture*, New York: Rockefeller Foundation.
Rogers, C.B. (1979) 'Poultry and eggs', in L.P. Schertz *et al.* (eds) *Another Revolution in US Farming?*, Washington, DC: US Department of Agriculture.
Rooth, T. (1985) 'Trade agreements and the evolution of British agriculture in the 1930s', *The Agricultural History Review* 33 (part II).
Rosenberg, C.E. (1977) 'Rationalization and reality in the shaping of American agricultural research, 1875–1914', *Social Studies of Science* 7.
Rossiter, M.W. (1979) 'The organization of the agricultural sciences', in A. Oleson and J. Koss (eds) *The Organization of Knowledge in Modern America, 1860–1920*, Baltimore: Johns Hopkins University Press.
Rowntree, B. S. (1901) *Poverty: A Study of Town Life*, London.

Royal Commission on Environmental Pollution (1989) Thirteenth Report: *The Release of Genetically Engineered Organisms to the Environment*, London: Her Majesty's Stationery Office.

Runciman, G. (1966) *Relative Deprivation and Social Justice*, London: Routledge & Kegan Paul.

Ruston, A.G. and Witney, D. (1934) *Hooten Pagnell*, London: Edward Arnold.

Sagoff, M. (1988) 'Biotechnology and the environment: what is at risk?', *Agriculture and Human Values* 5 (3), Summer.

Sahai, O., Knuth, M., Park, G., Nieder, M. and Yang, J.-H. (1986) 'Producing high value food ingredients via plant biotechnologies', *The World Biotech Report 1986*, vol. 2 (1): *Food Processing*, New York: Online International.

Sanderson, S.E. (1986) *The Transformation of Mexican Agriculture*, Princeton: Princeton University Press.

Saville, J. (1957) *Rural Depopulation in England and Wales*, London: Routledge & Kegan Paul.

Schramm, G. and Warford, J. (1989) (eds) *Environmental Management and Economic Development*, Baltimore: Johns Hopkins University Press/World Bank.

Seddon, D. (1986) 'Bread riots in North Africa: economic policy and social unrest in Tunisia and Morocco', in P. Laurence (ed.) *World Recession and the Food Crisis in Africa*, London: James Currey.

Senior, S. (1989) 'European interest groups and the CAP', *Food Policy* 14 (2).

Shackley, S. J. (1989) 'Regulation of the release of genetically manipulated organisms into the environment', *Science and Public Policy* 16 (4), August.

Shanin, T. (ed.) (1972) *Peasants and Peasant Societies*, Harmondsworth: Penguin.

Sharpe, S. (1984) *The Lives of Working Mothers*, Harmondsworth: Penguin.

Shaw, J.M. (ed.) (1979) *Rural Deprivation and Planning*, Norwich: Geo Books.

Shiva, V. (1989) *Staying Alive: Women, Ecology and Development*, London: Zed Press.

Silva, J. Souza (1988) 'The contradictions of the biorevolution for the development of agriculture in the Third World: biotechnology and capitalist interests', *Agriculture and Human Values* 5 (3), Summer.

Simpson, R. (1989) 'Feeding poverty', *New Statesman and Society*, 8 December.

Singh, R.B. (1989) *Current Status and Future Prospects of Plant Biotechnologies in Developing Countries of Asia*, paper presented at the CTA/FAO Symposium on Plant Biotechnology, Luxembourg.

Smart, B. (1985) *Michel Foucault*, London: Ellis Horwood & Tavistock.

Smith, N. (1984) *Uneven Development*, Oxford: Blackwell.

Teubal, M. (1987) 'Internationalization of capital and agroindustrial complexes: their impact on Latin American agriculture', *Latin American Perspectives* 14 (3), Summer.

Thompson, F.M.L. (1963) *English Landed Society in the Nineteenth Century*, London: Routledge & Kegan Paul.

Townshend, P. (1979) *Poverty in the United Kingdom*, Harmondsworth: Penguin.

Tracy, M. (1982) *Agriculture in Western Europe*, 2nd edn, London: Granada.

Traill, B. (1983) 'Simulating changes in investment grants, interest and tax rates and inflationary pressures on input prices in UK agriculture', *Journal of Agricultural Economics* XXXIV.

Tubiana, L. (1989) 'World trade in agricultural products: from global regulation to market fragmentation', in D.E. Goodman and M.R. Redclift (eds) *The International Farm Crisis*, London: Macmillan.

UNCTC (1981) *Transnational Corporations in Food and Beverage Processing*, New York: United Nations Centre on Transnational Corporations.

USDA (1989) *Rural and Rural Farm Population, 1989*, Washington, DC: US Department of Agriculture, Economic Research Service.

Vogtmann, H. (1985) 'Environmental and socio-economic aspects of different farming practices', *The Other Economic Summit*, London.

Walker, A. (1989) 'Europe steals the lead in plant biotechnology', *Biotech 89 Proceedings*, Pinner, Middlesex: Blenheim Online.

Weber, M. (1970) 'Capitalism and rural society in Germany', in H. Gerth and C. Wright Mills (eds) *From Max Weber*, London: Routledge & Kegan Paul.

Webster, A.J. (1989) 'Privatisation of public sector research: the case of a plant breeding institute', *Science and Public Policy* 16 (4): 224–32.

Weiskel, T.C. (1989) 'The ecological lessons of the past: an anthropology of environmental decline', *The Ecologist* 19 (3).

Westwood, S. (1984) *All day Every Day: Factory and Family in the Making of Women's Lives*, London: Pluto Press.

Whatmore, S. (1988) 'From women's roles to gender relations: developing perspectives in the analysis of farm women', *Sociologia Ruralis* xxviii (4): 239–48.

Whitelegg, E. *et al.* (1982) *The Changing Experience of Women*, Oxford: Martin Robertson/The Open University.

Wilken, R. (1987) *Good farmers*, Berkeley, Calif.: University of California Press.

Wilkinson, J. (1987) *Europe Within the World Food System: Biotechnologies and New Strategic Options*, FAST Exploratory Dossier 11, Brussels, Commission of the European Communities.

World Conservation Strategy (1983) *The British Government's Response to the WCS*, London: World Wildlife Fund.

Yeandle, S. (1984) *Women's Working Lives*, London: Tavistock.

Yotopolous, P. (1985) 'Middle-income classes and food crises: the ''new'' food-feed competition', *Economic Development and Cultural Change* 33 (3), April.

Zurek, E. (1986) Part-time farming in the Federal Republic of Germany', *Sociologia Ruralis* xxvi (3/4): 377–84.

271

INDEX

accumulation 43, 87–102;
   extensive 87, 93–9; intensive
   (fordist) 87–8, 89, 99–102,
   109, 134, 251–2; (in Third
   World 137–8, 163–4, 253–4)
acid rain 206
additives 244, 246,
advertising and marketing 4–5,
   39; and consumer resistance
   240–1, 244, 245; of
   convenience foods 37; of
   'healthy' foods 38, 240–1,
   244–5; of processed foods
   34, 36; and product
   differentiation 90, 240–1,
   244; of white goods 18, 39;
   and women 44, 244
Africa 153, 154, 155–6
Aglietta, M. 93–4
Agricultural Act (1947) 121
Agricultural Act (1954) (US)
   123
Agricultural Adjustment
   Administration (US) 117–18
Agricultural and Farming
   Research Council 117–18
Agricultural Genetics
   Company 177
Agricultural Holding Act
   (1923) 64–5
Agricultural Marketing Acts
   (1931 & 1933) 64
Agricultural Research Service
   (US) 179

Agricultural Trade
   Development and
   Assistance Act (1954) (US)
   see PL480 programme
Agriculture Act (1986) 221
Allen, J. 204
Amazonian rain forest 162,
   254–5
American Soybean Association
   107, 108–9
Anderson, P. 61
appliances see white goods
appropriation 90–1, 201; public
   perception of 243–4
Argentina 141, 158, 159

Baker, Nicholson: *The*
   Mezzanine 239
Barbier, E. 235–6
Barton, J.H. 182
Baxter, S. 25–6
beef production, Third World
   157–8, 160–2
Belasco, W.J. 37, 240, 243
Berlan, J.-P. 103–4, 108
Biosphere Reserves 228
biotechnology 167–71, 250;
   animal 111, 145, 169, 170,
   182; environmental concern
   about 168, 170, 195–7, 217;
   and North–South division
   180–1, 187, 188–94; plant 97,
   98, 101, 103–6, 115, 169,
   170; (privatization of

175–80; property rights in
181–7, 192; in Third World
145, 150–2, 168, 170;
transfer of technology
192–4; and transnational
corporations 171–2, 174,
194); research and
development in 168, 170,
171–80; (and ownership
180–7, 192)
biscuits 34, 37
Bishop, J. 204
Bose, C. 17–19
Bourdieu, Pierre 247, 251
Bradley, A. 80, 82, 83
Brazil 141, 142, 143, 144–7,
163–4; biotechnology in 193;
rainforest in 162, 254–5;
transnational corporations
in 145–6, 158, 160, 162
bread 35
British Restaurants 24–5
Bunker, S. 234
Burch, W.B, 230
Burns, J. 11
Busch, L. 189, 193
Buttel, F.H. 176, 180, 193, 194,
221

Cairns Club 128, 129
Calgene 171, 172, 174
canned food 30, 38
canteens 25
capital: and farm enterprise
71–5, 85–6; see also
accumulation
catering industry 24–6, 45;
employment in 25–6, 27–8,
40; fordism and 101; and
Third World 146, 158
Central America 151–2, 157–8,
159, 160–1, 162–3, 192;
resource management in 233
cereal(s): consumption 34–5,
37, 40; production 59–60
Chambers, R. 149
Charles, N. 248–9
Charvet, J.-P. 157
chemicals: industry 97–8; (and

biotechnology 171–5);
resistance to 170, 171, 172;
see also fertilizers
Cherfas, J. 169
Cheshire, P. 208, 211
Ciba-Geigy 172, 174
coal mining 61
Cochrane, W.W. 112–14,
123–4, 134, 139
Cockburn, Cynthia 16–17
Collins, E. 34–5
Colombia 143–4
Commission on Plant Genetic
Resources (CPGR) 186–7
Commodity Credit Corporation
(US) 117–18
Common Agricultural Policy
(CAP) (EC) 122, 127,
129–30, 136–7, 206–10
community, notion of 77, 78,
80, 81–2, 86
Connor, S. 177
Consultative Group on
International Agricultural
Research (CGIAR) 152, 184,
185–6, 192, 214
consumption 4–7, 29–36; and
extensive accumulation 93–4;
and fordism 99, 109, 137–8,
163–4; post-fordist era of
240–1; recent changes in
28–9, 36–43; (patterns of
resistance 42–3, 130, 240–6,
252–3; in Third World 155–7,
163–4); sociology of 246–51;
uniformity of 158; and white
goods 22–3; see also diet
convenience foods 29, 37–8,
39, 40; and convenient
shopping 40–1; fordist
consumption of 99; and
Third World 158, 160–2;
and women 10–11, 37–8,
248–9; see also fast food
Coolidge, Calvin 116
corn: high fructose syrup from
190; hybrid seeds of 103–4,
105–6, 145, 169
cotton 104–5

Council for the Protection of
Rural England 218
countryside *see* rural areas
Countryside Commission 219
crop yields: in Latin America
143; in United States 98-9,
105-6, 110

Dade, P. 244
Darling, George 24
debt, Third World 144, 154,
228-9, 255
de Janvry, A. 137, 142
Deo, S.D. 192
deprivation, social 79-83, 86,
141-2, 147, 215, 254
developing countries *see* Third
World
Diamond v. Chakrabarty 181
diet 31-2; and processed food
33-6; recent changes in
37-8, 40, 240-6, 252-3; in
Third World 155-7, 253-4,
255; in United States 109,
245-6; *see also* nutrition
distribution industry, food:
employment in 26, 27; in
Third World 146
Douglas, Mary 247
Dumont, René 53, 54
DuPont 172, 174

'ecologism' 222-3
economy 6-7; and
environment 233-6;
household 14-16, 29-30, 31,
32-3, 36-7, 38-9
Eisenhower, Dwight D. 123
employment: in agriculture 27,
28, 49-53, 54, 59; (in Brazil
146; in Europe 68, 69-71);
in catering industry 25-6,
27-8; in food industry 26-8,
90; of women *see under*
women
energy transfer 234
Engels, Friedrich 6
environment(al): concerns 49,
89, 91, 130, 201-6, 210-17;

(over biotechnology 168,
170, 195-7, 217; in Europe
207, 208-9, 210-13, 222;
in Third World 161-2,
213-16, 223-6); consciousness of 230-2; management
217-29, 252, 254-5; and
'naturalness' 250-1;
political movement 223,
225-6, 235-8; and society
229-36
Europe/European Community
(EC), agriculture in 66-71,
84, 85, 95-7, 153; and
biotechnology 174, 176-8,
182; Common Agricultural
Policy 122, 127, 129-30,
136-7, 206-10; conflict with
US over 127, 128, 129;
environmental concern over
207, 208-9, 210-13, 222;
Marshall Plan 135-6; state
intervention in 119-22, 216;
technology/policy model
102-3, 209; usable area for
206
European Agricultural
Guidance and Guarantee
Fund 207
European Commission 182,
252
Eversley, Lord 52-3
Exmoor National Park 219
expenditure, household 38-9;
on food 29-30, 31, 32-3,
36-7, 39

family farming 47-9, 53-9, 65,
83-6; and capital 73-5, 92;
in Europe 67-71; in Third
World 138, 141-2, 146-7,
148-9; women's role in 1-2,
45-6
Farming Systems Research 228
fast food 25-6, 45; and fordism
101; technological changes
in 28; and Third World 146,
158, 160-2; trivialization of
239

Federal Food and Drug Administration (US) 246
feeds, animal 107, 108–9, 110; and biotechnology 189
fermentation biotechnologies 188, 189
fertilizers 61, 62, 211–12; in Latin America 143, 145; and plant genetics 103, 104
Feyerabend, Paul 231
Finegold, K. 117
Fite, G.C. 104–5, 117–18
Food and Agriculture Organization 186–7
Food for Peace Act (1966) (US) 139
'Food for Peace' (PL480) programme (US) 123–4, 126, 139, 154
fordism 99; and agriculture 87–8, 89, 100–2, 109, 134, 251–2; in Third World 137–8, 163–4, 253–4
foreign exchange 139, 140, 229
forests: conversion of 212–13; tropical 162, 228–9, 254–5
fossil fuels 112–14
Fowler, C. 193
freezers 19–20, 21, 22, 30, 33
fridges 19–20, 21, 33, 39
Friedmann, H. 135, 136, 154

GATT (General Agreement on Tariffs and Trade) 135; Uruguay Round of negotiations 128–9, 183, 186
gender divisions 3–4; see also women
genetic engineering see biotechnology
Gershuny, J.I. 247
Giddens, A. 230, 248
Gofton, L. 246–7
Goodman, D.E. 150, 170, 173, 190, 191
Goody, Jack 247
Goonatilake, S. 230–1
Gorz, André 222–3
government(s) 48, 54–5, 76, 84, 85, 93, 114–30; and biotechnology 177–8, 179–80; and consumer resistance 245–6; and environmental problems 214, 215, 216–7, 219–21, 223, 228–9, 252; and land ownership 62, 64–5; in United States 101, 102, 107–9, 122–9; (New Deal policies 88, 109, 115–19, 122–3, 133).
grain: fordism and 102; in United States 98, 103–4, 105–6, 110; world market in 94–7, 126, 147, 153–5
Green Revolution 142–4, 148–9, 150–2, 192, 214–15

Hadwiger, D.F. 118
'hamburger connection' 158, 160–2
Heady, E.O. 106
'healthy' foods 38, 40, 240–1, 242–3, 244–5, 252–3
hedgerows, removal of 210–11, 219
Hibberd (Ex parte Hibberd) 182, 186
high fructose corn syrup 190
Hobsbawm, Eric 95, 96
Hoechst 173, 174
Hooten Pagnell 55–9, 203
household composition 39
housework 8–12, 44–5, 242; *versus* employment 14–16, 23; on family farms 75; see also white goods
Howes, M. 231
hybrid seeds 97, 98, 101, 103–6, 115, 169, 194; property rights to 182–3; in Third World 145, 150–2

ICI 172, 173, 174
Indemnité Viagière de Départ policy (France) 69
Indonesia 204, 216
industrialization 4, 6–7; of

agri-food system 89, 90–3;
in Third World 139–40, 142
innovation in agri-food system
89, 90–3
Integrated Pest Management
practices 216
Integrated Rural Development
Programmes 147–9
International Agricultural
Research Centres (IARCs)
152, 184, 192, 214, 215
International Bank for
Reconstruction and
Development 135
International Board of Plant
Genetic Resources (IBPGR)
184, 185, 187
International Centre for
Genetic Engineering and
Biotechnology 193
International Convention for
the Protection of New Plant
Varieties 181
International Monetary Fund
(IMF) 135, 144
International Undertaking on
Plant Genetic Resources
186–7

Jackson, Robert 177–8
Jenkins, D. 22–3
Jiggins, J. 149
Johnson, Lyndon B. 124
Johnston, J. 10
Juma, C. 182, 184, 185

Kennedy, John F. 124
Kenney, M. 179
Kerr, M. 248–9
Kidd, G. 172
Kindleberger, C.P. 97, 115
kitchens, fitted 19–20
Kloppenburg, J. Jr 104, 179,
182, 185
Komer, D.I. 161
Kornberg, Hans 196

labour see employment
Lacy, W.B. 193

land: conversion 212–13;
ownership 73–4; (in Britain
51, 55, 56, 58–9, 60–5, 85; in
Europe 67–8, 85; in Latin
America 141); use 29, 53,
56–7, 84, 205–6; (in Latin
America 141, 142–4); values
209, 210–11
Land Grant University system
(US) 176, 179
Latin America 139, 140–7, 152,
158, 159–60, 162, 163–4;
grain trade by 153, 154; rain
forest of 162, 254–5
Lewontin, R.C. 103–4
livestock: in Central and Latin
America 145, 158, 160–2;
feedstuffs for 107, 108–9,
110, 189; in United States
107, 108–9, 110–12
Lockeretz, W. 108, 109
London Food Commission 243
Luke, T. 232–3

McDowell, L. 14–15
McGovern Report (US) 243,
245–6
Mali 204
Malthusian population theory
235
Mansholt Plan (1968) 68
manufacturing industry, food
7, 28–9, 97; and consumer
resistance 240–1, 243–6;
employment in 26, 27, 28,
90; fordism in 101–2; in
Third World 146, 150, 158
marketing see advertising
Marshall, Alfred 235
Marshall Plan 135–6
Massey, D. 250
Mathias, P. 36
meat, US consumption of
111–12, 245–6; Third World
and 157–8, 160–2
Meissner, M. 23
Mendras, Henri 66
Mexico 151, 157–8, 159, 162–3,
192

Microbiological Resource
Centres 193
microwave ovens 19–22, 39,
253
Milk Marketing Board 54, 120
milk processing industry, Latin
American 159
Mingione, E. 15–16
Mitterand, François 66–7
Mond, Sir A. 55
Monsanto 172, 173, 174
multinational corporations
145–6, 150, 157–62, 170,
171–5, 190, 194

Nash, R. 226
National Fruit Trials station,
Brogdale 178
National Parks 219, 226
National Seed Development
Organization 178
Nations, J.D. 161
'natural' concept of 250–1
Nature Conservancy Council
(NCC) 218, 219, 220
Neighbourhood Watch
schemes 78
New Deal (US) 88, 109,
115–19, 122–3, 133
Newby, Howard 76
nitrate pollution 212
nitrogen economy 97–8
Nixon, Richard M. 124, 125,
147
Nordskog, A.W. 111
North–South division 253–6; in
biotechnology 180–1, 187,
188–91; and environmental
management 225–9, 254–5
nutrition: in Britain 31–2; in
Third World 141, 144, 155,
163

O'Day, R. 13
organic food 243–4

Pahl, J. 77, 79, 247
part-time employment: in
farming 69–70, 217; of

women 27
patenting of genetic resources
180–4, 186, 192
Peoria Plan (1928) 107
Peru 159–60
pesticide residues in food 243
pesticide subsidies 216
PL 480 programme (US) 123–4,
126, 139, 154
plant breeding see under
biotechnology
Plant Breeding Institute,
Cambridge 172, 178
Plant Variety Protection Act
(PVPA) 1970 (US) 181–2
Platteau, J.-P. 155
Poor Laws 51
population growth 234–5; in
cities 49–53, 156–7
post-fordist consumption 240–1
Potter, C. 218, 219, 222
poultry production, US 111
preservation, food 29
prices of food: changes in
32–3, 111; support policies
for 76, 88, 100, 109, 126,
134; (during New Deal 116,
117, 118, 122–3; see also
Common Agricultural
Policy); in Third World
137–8
processing/processed food 30,
33–6, 97; and biotechnology
189–91, 250; and consumer
resistance 241, 243–4;
expenditure on 36–7;
'naturalness' of 250;
substitution and 91–2,
243–4; in Third World 146,
158–9, 161; women's work
in 12–14; see also
convenience foods; fast food
Public Health Act (1872) 49–50

Rama, R. 109, 157–8
Rappaport, R. 233
Rasmussen, W.D. 98
rationing 24, 121
Raw, G. 25–6

INDEX

Reeves, Margaret 31
Repetto, R. 204
research and development,
    biotechnological 168, 170,
    171–80; ownership of results
    of 180–7, 192
restaurants 24–5, 45
retailing 28, 34, 36, 40–3;
    employment in 26, 27, 28
Rhone-Poulenc 172, 174
Ricardo, David 235
Rice, Marjorie Spring 8–9
rice production, Colombian
    143–4
Roosevelt, Franklin D. 109, 117
Rooth, T. 120
Rowntree, B.S. 31–2
rural areas 48–9, 76–83;
    definition of 49–50, 79;
    employment in 27, 28,
    49–53; Integrated
    Development Programmes
    147–9; social deprivation in
    79–83, 86; (in Third World
    141–2, 147, 215, 254); see
    also family farming
Ruston, A.G. 55–9
Ryan, M.E. 112–14, 123–4, 134,
    139

Sadoulet, E. 142
Sagoff, M. 170
Sanderson, S.E. 158, 163, 164
Saville, John 49
seed firms 171–2, 174
seeds see hybrid seeds
Senior, S. 208
servants, domestic 7–8, 53
Settlement Laws 51
shopping for food see retailing
Silva, J. Souza 192
Simpson, R. 207
slurry, disposal of 206, 212
Soil Bank programme (US) 124
soil erosion 204
Sorj, B. 150
soya beans: in Latin America
    141, 146, 147, 164; in United
    States 107–9, 126, 136–7

Staley Company 107
state, role of see government
Structure Plans 76
substitution 91–2, 101, 202;
    and biotechnology 188–91;
    and environment 236;
    public perception of 243–4
sugar 35, 37; alternatives to
    190
supplements, 'healthy' food
    244–5
surpluses: in Europe 127,
    129–30; in United States
    123–4, 134–5, 138–9, 154
Swanson, L.E. 192

tea 35–6
technology 1,2, 52, 73–4, 92–3,
    97–9; Green Revolution
    142–4, 148–9, 150–2, 192,
    214–15; technology/policy
    model 102–6, 111, 112–14;
    (internationalization of
    133–65, 209, 215); see also
    biotechnology; white goods
Teubal, M. 155
Third World 253–6; and
    biotechnology 145, 150–2,
    168, 170, 183–7, 191–4;
    (North–South division
    180–1, 187, 188–91); debt
    crisis of 144, 154, 228–9,
    255; environmental
    problems of 161–2, 213–16,
    223–6, 254–5; (governments
    and 215, 216–17, 254); food
    aid to 123, 126, 134–5,
    138–9; food crisis in 140,
    144, 153, 162–4; (in Africa
    155–6, 157); technology
    transfer to 191–4;
    ('environment-friendly' 228;
    technology/policy model
    103, 112, 137–64, 215);
    women in 2–3
tissue culture, industrial 188,
    189–90
Trail, B. 209
transnational corporations

278

145–6, 150, 157–62; and
biotechnology 170, 171–5,
190, 194
Tubiana, L. 136

Unilever 172, 174
United States 68, 83–4, 85,
94–9; biotechnology in 173,
174, 176, 179–80, 181–3;
consumer resistance in 240,
243, 245–6; environmental
management in 221; and
fordism 100–2; and soil
erosion 204; state
intervention in 101, 102,
107–9, 122–9; (biotechnology
179–80; conflict with Europe
127, 128, 129; New Deal
policies 88, 109, 115–19,
122–3, 133);
technology/policy model
102–14; (internationalization
of 133–65, 209, 215)
universities: and genetic
research 173, 176–7
urbanization 49–53, 156–7
Uruguay Round of GATT
negotiations 128–9, 183, 186

Vavilov centres 168, 184
video recorders 22
vitamin and mineral
supplements 244–5

Wallace, C. 247
Wallace, Henry A. 151
War, Second World 23–5, 30,
48, 59, 120–1, 122–3
waste: animal 206, 212; toxic
206
Webster, A.J. 178
Weiskel, T.C. 229

wetlands, conversion of 212–13
Whatmore, Sarah 74–5
white goods 11, 15, 16–23, 30,
32, 33, 44–5; expenditure on
39; fordist consumption of
99, 109; and 'healthy' foods
253
Wildlife and Countryside Act
(1981) 220
Wilken, R. 234
Wilkinson, J. 150, 170, 190, 191
Winrock Report (US) 179
Witney, D. 55–9
women 1–5, 44, 242; diet and
nutrition of 31–2; and
expenditure on food 30;
labour by 1–2, 5, 10–14,
44–5, 90; (as agricultural
labourers 52–3; in catering
industry 25, 27–8; and
convenience foods 10–11,
37–8; as domestic servants
7–8, 53; on family farm 1–2,
45–6, 70–1, 75, 84; in food
processing 12–14; and
fordist accumulation 99; in
home 8–12, 14–16, 23, 44–5,
75, 242; *versus* housework
14–16, 23; naturalization of
250)
Woolton, Lord 24
world market 60, 94–7; and
technology/policy model
133–65, 209, 215
World Bank 216
World Conservation Strategy
225–6
World Intellectual Property
Organization 183

Yeandle, S. 15
Yotopoulos, P. 155
Young, Arthur 55